LIBRARY
HARPER ADAMS UNIVERSITY COLLEGE

69090

LONG LOAN

RETURNED

2 8 OCT 2004

**THIS ITEM MUST BE RETURNED OR RENEWED BY THE LAST
DATE STAMPED ABOVE.
FINES ARE PAYABLE FOR LATE RETURN OF ITEMS.**

D1639691

WITHDRAWN

Worker Exposure to Agrochemicals

Methods for Monitoring and Assessment

Edited by
R.C. Honeycutt
Edgar W. Day, Jr.

LEWIS PUBLISHERS

Boca Raton London New York Washington, D.C.

WITHDRAWN

HARPER ADAMS UNIVERSITY LIBRARY COLLEGE

69090

Library of Congress Cataloging-in-Publication Data

Worker exposure to agrochemicals : methods for monitoring and assessment /
R.C. Honeycutt, Edgar W. Day, Jr., editors
    p. cm.
  Includes bibliographical references and index.
  ISBN 1-56670-455-3 (alk. paper)
    1. Pesticides—Toxicology. 2. Agricultural laborers—Health risk assessment.
3. Biological monitoring. 4. Pesticides—Safety measures.
I. Honeycutt, Richard C., 1945–. II. Day, Edgar W.

RA1270.P4 W59 2000
615′.902—dc21
                                           00-057369

This book contains information obtained from authentic and highly regarded sources. Reprinted material is quoted with permission, and sources are indicated. A wide variety of references are listed. Reasonable efforts have been made to publish reliable data and information, but the author and the publisher cannot assume responsibility for the validity of all materials or for the consequences of their use.

Neither this book nor any part may be reproduced or transmitted in any form or by any means, electronic or mechanical, including photocopying, microfilming, and recording, or by any information storage or retrieval system, without prior permission in writing from the publisher.

All rights reserved. Authorization to photocopy items for internal or personal use, or the personal or internal use of specific clients, may be granted by Lewis Publishers, provided that $.50 per page photocopied is paid directly to Copyright Clearance Center, 222 Rosewood Drive, Danvers, MA 01923 USA. The fee code for users of the Transactional Reporting Service is ISBN 1-56670-455-3/01/$0.00+$.50. The fee is subject to change without notice. For organizations that have been granted a photocopy license by the CCC, a separate system of payment has been arranged.

The consent of Lewis Publishers does not extend to copying for general distribution, for promotion, for creating new works, or for resale. Specific permission must be obtained in writing from Lewis Publishers for such copying.

Direct all inquiries to Lewis Publishers, 2000 N.W. Corporate Blvd., Boca Raton, Florida 33431.

Trademark Notice: Product or corporate names may be trademarks or registered trademarks, and are used only for identification and explanation, without intent to infringe.

© 2001 by CRC Press LLC
Lewis Publishers is an imprint of CRC Press LLC

No claim to original U.S. Government works
International Standard Book Number 1-56670-455-3
Library of Congress Card Number 00-057369
Printed in the United States of America 1 2 3 4 5 6 7 8 9 0
Printed on acid-free paper

Foreword

R.C. Honeycutt, Ph.D.

Methods to determine exposure of pesticides to agricultural workers have developed over the last 30 years at a reasonably slow pace. Some of the first methods to determine agricultural worker exposure to pesticides were developed by Armstrong, Wolfe, and Durham[1] as well as Poppendorf.[2] Using these methods for research has resulted in the development of several axioms related to agricultural worker exposure to pesticides:

1. In most cases, dermal exposure is much more of a significant route of worker exposure than respiratory exposure.
2. The hands of workers who do not wear gloves while working with pesticides are usually the most contaminated part of the body when workers perform such chores as mixing and loading of the pesticides or picking commodities during re-entry in pesticide-treated crops.
3. The patch method (although not as popular today) and the use of whole-body dosimeters are both reasonable methods for determining dermal deposition on the skin of agricultural workers during application of pesticides or when they re-enter treated fields.
4. Generally speaking, the equipment that workers use and the type of agricultural work performed while handling pesticides or pesticide-treated commodities have a major impact on the level of exposure a worker receives. Neat work habits generally lead to lower exposure for agricultural workers who handle pesticides. The current Pesticide Handlers Exposure Database (PHED), which is used to predict exposure to mixers/loaders/applicators, is based on this concept.

One might now ask what further advances are necessary to adequately understand and determine agricultural worker exposure. This is a fair question. Even with the above axioms in place, there still remains a great debate over the use of dermal deposition techniques as opposed to the use of biological monitoring to measure agricultural worker exposure. These divergent techniques are said to be incompatible, and very few researchers have been able to show the same results in concurrent experiments using the two techniques. Another area that requires significant research is agricultural re-entry exposure — how to adequately measure it and predict it using foliar dislodgeable residue techniques.

This book is a compilation of research papers presented at the 1996 spring meeting of the American Chemical Society in New Orleans, LA.

Biological monitoring vs. dermal dosimetery methods to measure worker exposure are explored, and the simultaneous use of the two techniques is reviewed. Methods adopted by the industry for the measurement of foliar dislodgeable residues on crops are also discussed in this book. Although the information presented at the 1996 ACS meeting has been public for some time, the details of the techniques for measuring worker exposure in this book are relatively new and exciting.

Chapter one is a review of various classes of pesticides and how biological monitoring can be used to determine environmental concentrations of agrochemicals. Chapters two and four deal with methods to simultaneously measure exposure of individuals to agrochemicals using passive dosimetry in conjunction with biological monitoring. Chapter three is an excellent presentation that describes how a risk assessment for farm workers may be carried out using exposure data from biological monitoring. Chapter five explores the use of protective clothing to reduce worker exposure and relates how biological monitoring can be used as a tool in this regard. Chapter six explores the differences in the methods used for whole-body dosimetery in Canada vs. those used in other countries.

Chapter seven of this book describes the technique of using Jazzercise™ to measure pesticide residues on persons who may re-enter a pesticide-treated residence. Chapter eight describes a tiered approach to estimating exposure to workers who re-enter pesticide-treated crops. Chapter nine deals with modeling re-entry exposure, and Chapter ten describes techniques to perform a dislodgeable residue study and to estimate or model re-entry exposure from pesticide residues on treated turf. The last chapter of this book describes Good Laboratory Practice (GLP) Standards requirements necessary to perform worker exposure studies. GLP requirements are standard in today's agrochemical research, and, although some may argue that this chapter is out of place or may not belong in this book, it could be said that this is one of the most important chapters in this book and should be reviewed carefully in order to understand the GLP requirements one may face when planning or designing such complex field studies as described in the research chapters of this book.

There are some research issues that remain to be resolved in the area of agrochemical farmworker exposure assessment, including: (1) the concept of modeling worker exposure and how modeling may be applied to re-entry exposure assessment; (2) the accuracy of exposure data generated in worker exposure studies and how to deal with the imprecision inherent in these exposure estimates; and (3) education of farmworkers to minimize exposure during mixing and loading or re-entry activities. It is well known that exposure can be mitigated by wearing protective clothing, such as a long-sleeved shirts, pants, boots, and gloves, while working, as well as working in a tidy fashion.

I would like to thank all the contributors to this book who gave up their valuable time to see that the work described herein would be published. The new and advanced techniques for estimating agricultural worker exposure and risk described in this book have been developed over the past several years, and the information on these techniques will remain relevant for some time to come. This book certainly adds to the knowledge available to researchers who deal with agrochemical worker exposure issues and will be valuable as a reference tool to future researchers.

References

1. Wolfe, H.R., Durham, W.F., and Armstrong, J.F., *Archives of Environmental Health*, 14, 622, 1967.
2. Poppendorf, W.J., Advances in the unified field model for reentry hazards, in *Dermal Exposure Related to Pesticide Use: Discussion of Risk Assessment*, Honeycutt, R.C., Zweig, G., and Ragsdale, N., Eds., ACS Symposium Series 273, American Chemical Society, Washington, D.C., 1985.

Introduction

Edgar W. Day, Jr.

Historical review of sampling methods

Soon after the organophosphate insecticides were first introduced in the late 1940s, episodes of illness in workers associated with the handling of these products began to be reported. This led to an interest in ascertaining the level of exposure to these chemicals that workers might experience while performing their normal work tasks.

The first pesticide exposure study was reported by Griffiths et al. (1951). Parathion was trapped on respirator filter discs during application to citrus trees. Batchelor and Walker (1954) expanded exposure monitoring to include the estimation of potential dermal exposure using pads attached to workers' clothing. Durham and Wolfe (1962), in their classic review of worker exposure methodologies, also provided some experimental validation for the best available methods.

From the late 1960s until the early 1980s, a large number of worker exposure studies were reported which used the methods of passive dosimetry — that is, methods that measured potential contact with pesticides but did not measure the actual amount of pesticide absorbed by the workers' bodies. These studies were extensively reviewed by Wolfe (1976) and later by Davis (1980).

The most common methods for estimating respiratory exposure during this time period involved the use of a personal air-sampling system or the insertion of an adsorbing material inside a respirator. The latter technique was not widely used because tight-fitting respirators were difficult to obtain and were uncomfortable for field workers to use in the field. The personal sampler generally consisted of a small air pump that pulled air through a tube containing an adsorbent or other material to trap the airborne pesticide. The collection tube was placed near the breathing zone of the worker or at a nearby fixed location if the sampler was too restrictive for the worker. The potential exposure was calculated from the amount of pesticide collected, the volume of air pulled through the tube, and standard inhalation rates for workers. This methodology is essentially still used in the field, the main advances being in the variety of adsorbent materials and the use of filter cassettes instead of tubes.

Through the early 1980s, dermal exposure to the body (except hands) was measured via some type of patch that varied in size and composition. In early studies, patches were placed only on exposed parts of the body (arms, neck, head). Later studies included patches to represent all body parts, both inside and outside of normal clothing. Hand exposure, which makes an extremely important contribution to overall dermal exposure, was generally monitored using lightweight, absorbent gloves or by swabbing or rinsing the hands with various solvents in a variety of ways. Both the respiratory and dermal exposure techniques were well described by the U.S. Environmental Protection Agency in Subdivision U of their Pesticide Guidelines (1986).

In the early 1980s, the whole-body dosimeter (WBD) was introduced as a superior method for passive dermal dosimetry monitoring. A standard protocol was described by the World Health Organization (1982), and Abbott et al. (1987) described some additional options. Chester (1993) reported refinements that permitted exposure estimation by passive dermal dosimetry and biological monitoring simultaneously.

In the mid-1980s, Fenske et al. (1986a,b; 1993) described the use of fluorescent compounds coupled with video-imaging measurements to produce exposure estimates over virtually the entire body. The technique has been applied to pesticide mixers and applicators by Fenske and his coworkers, but has not been widely used by other investigators.

Biological monitoring is believed by most exposure and risk assessment scientists to provide the most accurate measurement of exposed dose to pesticides for field workers when the method can be utilized. An early example of the use of biomonitoring is the analysis of cholinesterase levels in blood which have been used as indicators of worker exposure to cholinesterase-inhibiting chemicals such as organophosphate and carbamate pesticides (Peoples and Knaak, 1982). Urinary metabolites have been used to detect exposure to pesticides during field operations since the late 1960s (Davies et al., 1979; Durham and Wolfe, 1962; Levy et al., 1981; Swan, 1969). However, in order to quantitate exposure using biomonitoring, the pharmacokinetics of the chemical of interest must be known and fully understood so that the appropriate tissue, fluid, or excretion pathway, as well as appropriate time periods for monitoring, can be selected. Chester (1993) and Woollen (1993) have discussed the most appropriate methods for biological monitoring.

Surrogate databases and task forces

A significant advance in the development of methods for measuring chemical exposure to agricultural workers was an indirect one — namely, the development of surrogate databases and the establishment of cooperative task forces for the development of the databases. In the early 1980s, a concept began to develop in which it was assumed that the level of worker exposure was more dependent on the physical processes of mixing, loading, and application of pesticides than on the nature of the chemical. This concept began to take form when Severn (1982) reported that the U.S. EPA was using existing published studies to make preliminary exposure assessments for pesticides that were used in similar agricultural situations. Hackathorn and Eberhart

(1985) summarized some data that had been generated by their company and suggested that there was likely a large amount of similar data in other company files which, if compiled into a database, would likely support the hypothesis that exposure levels were driven by physical parameters. Their proposal, along with information published by others (Honeycutt, 1985; Reinert and Severn, 1985), ultimately led to the development of the Pesticide Handlers Exposure Database (PHED) in North America and the European Pesticide Operator Exposure Monitoring (EUROPOEM) database in Europe.

Both PHED and EUROPOEM are generic databases containing measured exposure data for workers involved in the handling or application of pesticides in the field. PHED was designed by a task force consisting of representatives from Health Canada (HC), the U.S. Environmental Protection Agency (U.S. EPA), and the American Crop Protection Association (ACPA). A similar group of industry and regulatory representatives was later responsible for the development of EUROPOEM. Versar, Inc., of Springfield, VA, developed the software for both systems. These databases allow exposure and risk assessments to be conducted with a larger degree of certainty, as exposure estimates are based on a much larger number of observations than those available from a single exposure study.

Establishment of the PHED and EUROPOEM databases has permitted the estimation of exposures and risks for a large number of pesticide use scenarios without the actual conduct of field studies. This has resulted in substantial savings of resources for the pesticide industry and the government regulatory community. Of equal or greater benefit was the demonstration by the PHED task force that industry representatives and pesticide regulators could work cooperatively toward goals of significant benefit to both parties. This led not only to a similar cooperative effort in Europe but also to the formation of other industry task forces that have operated with the full cooperation of the regulatory community. Of note are the Agricultural Re-entry Task Force (ARTF), whose purpose is to develop a generic database that will permit the estimation of exposure for persons who re-enter areas treated with pesticides, and the Outdoor Residential Exposure Task Force, which is developing sufficient data for potential databases that may be applied to residential situations.

A spin-off of all of these task forces has been the open discussions that have led to improved design considerations and effective use of resources in the conduct of field exposure studies. These task forces have evaluated a variety of exposure measuring techniques, developed study designs for conducting studies, and performed field studies in a uniform and efficient manner. The task force protocols and designs have become models for the industry, having received valuable input and approval from the regulatory community.

Among the more important design considerations of the task forces has been the lowering of detection limits of analytical methods to reasonable and attainable levels. These lower levels have resulted in lower default values (non-detectable levels) for use in risk assessments. Conversely, emphasis has been placed on the evaluation of the toxicity of a chemical and the utilization of such information in estimating required detection limits (default values) that would result in acceptable risk assessments.

Emphasis of this volume

The emphasis of the symposium upon which this book is based was on updating the literature of current methods used in exposure and risk assessment. Not all of the methods are currently acceptable for the regulation of pesticide chemicals; however, some may provide bases for future development of the science. The chapters of this book also put a strong emphasis on the value of study design, including the need for lower detectability. The final chapter by Hill and Swidersky emphasizes quality assurance, good laboratory practices, and worker safety in the conduct of studies. These are critical aspects of a well-designed study and, when included in the design, will likely result in the generation of valid and useful data.

Future considerations

The industry task forces (ARTF, ORETF, and others) are generating model protocols, efficient and accurate methods of sample collection, and analytical methods of appropriate detectability for use in field-worker exposure studies. Subsequently, the task forces are conducting field studies that will generate data for inclusion in several generic databases. It is understood that the databases will be the property of the member companies who have financed the work of the task forces. It is hoped, however, that the task forces will see fit to publish their protocols, methods, study designs, and other useful information in a volume like this one so that other scientists working in this discipline may access the information.

References

Abbott, I.M., Bonsall, J.C., Chester, G., Hart, T.B., and Turnbull, G.J. (1987) Worker exposure to a herbicide applied with ground sprayer in the United Kingdom, *Am. Ind. Hygiene Assoc. J.*, 48:167–175.

Batchelor, G.S. and Walker, K.C. (1954) Health hazards involved in the use of parathion in fruit orchards in north central Washington, *AMA Arch. Ind. Hygiene*, 10:522–529.

Chester, G. (1993) Evaluation of agricultural worker exposure to, and absorption of, pesticides, *Am. Occup. Hygiene*, 37:509–523.

Davies, J.E., Enos, H.F., Barquet, A., Morgade, C., and Danauskas, J.X. (1979) Developments in toxicology and environmental sciences: pesticide monitoring studies. The epidemiologic and toxicologic potential of urinary metabolites, in *Toxicology and Occupational Medicine*, Deichman, W.B., Ed., pp. 369–380.

Davis, J.E. (1980) Minimizing occupational exposure to pesticides: personal monitoring, *Residue Rev.*, 75:33–50.

Durham, W.F. and Wolfe, H.R. (1962) Measurement of the exposure of workers to pesticides, *Bull WHO*, 26:75–91.

Fenske, R.A. (1988) Correlation of fluorescent tracer measurements of dermal exposure and urinary metabolite excretion during occupational exposure to malathion, *Am. Ind. Hygiene Assoc. J.*, 49:438–444.

Fenske, R.A. (1990) Nonuniform dermal deposition patterns during occupational exposure to pesticides, *Arch. Environ. Contam. Toxicol.*, 19:332–227.

Fenske, R.A. (1993) Dermal exposure assessment technique, *Ann. Occup. Hygiene*, 37(6):687–706.

Fenske, R.A., Leffingwell, J.T., and Spear, R.C. (1986a) A video imaging technique for assessing dermal exposure. I. Instrument design and testing, *Am. Ind. Hygiene Assoc. J.*, 47:764–770.

Fenske, R.A., Wong, S.M., Leffingwell, J.T., and Spear, R.C. (1986b) A video imaging technique for assessing dermal exposure. II. Fluorescent tracer testing, *Am. Ind. Hygiene Assoc. J.*, 47:771–775.

Fenske, R.A., Birnbaum, S.G., Methner, M.M., and Soto, R. (1989) Methods for assessing fieldworker hand exposure to pesticides during peach harvesting, *Bull. Environ. Contam. Toxicol.*, 43:805–815.

Griffiths, J.T., Stearns, Jr., C.R., and Thompson, W.L. (1951) Parathion hazards encountered spraying citrus in Florida, *J. Econ. Entomol.*, 44:160–163.

Hackathorn, D.R. and Eberhart, D.C. (1985) *Database Proposal for Use in Predicting Mixer-Loader/Applicator Exposure*, ACS Symposium Series 273, American Chemical Society, Washington, D.C., pp. 341–355.

Honeycutt, R.C. (1985) *The Usefulness of Farm Worker Exposure Estimates Based on Generic Data*, ACS Symposium Series 273, American Chemical Society, Washington, D.C., pp. 369–375.

Levy, K.A., Brady, S.S., and Pfaffenberger, C.D. (1981) Chlorobenzilate residues in citrus worker urine, *Bull. Environ. Contam. Toxicol.*, 27(2):235–238.

Peoples, S.A. and Knaak, J.B. (1982) Monitoring pesticide/blood cholinesterase and analyzing blood and urine for pesticides and their metabolites, in *Pesticide Residue and Exposure*, Plimmer, J.R., Ed., ACS Symposium Series 182, American Chemical Society, Washington, D.C., pp. 41–57.

Reinert, J.C. and Severn, D.J. (1985) *Dermal Exposure to Pesticides: EPA's Viewpoint*, ACS Symposium Series 273, American Chemical Society, Washington, D.C., pp. 357–368.

Severn, D.J. (1982) Exposure assessment for agricultural chemicals, in *Genetic Toxicology, and Agricultural Perspective*, Fleck, R.A. and Hollander, A., Eds., Plenum Press, New York, pp. 235–242.

Swan, A.A.B. (1969) Exposure of spray operators to paraquat, *Brit. J. Ind. Med.*, 26:322–329.

U.S. EPA (1986) *Pesticide Assessment Guidelines. Subdivision U: Applicator Exposure Monitoring*, U.S. Environmental Protection Agency, Office of Pesticide Programs, Washington, D.C.

Wolfe, H.R. (1976) Field exposure to airborne pesticides, in *Air Pollution from Pesticides and Agricultural Processes*, Lee, Jr., R.E., Ed., CRC Press, Cleveland, OH, pp. 137–161.

Woollen, B.H. (1993) Biological monitoring for pesticide absorption, *Am. Occup. Hygiene*, 37:525–540.

WHO (1982) *Field Surveys of Exposure to Pesticides*, Standard Protocol VBC/82.1, World Health Organization, Geneva.

About the editors

Richard C. Honeycutt, Ph.D., was born in Newport News, VA, in 1945. He attended Anderson University in Anderson, IN, from 1963 to 1967 and earned an A.B. in Chemistry. He received his Ph.D. in Biochemistry from Purdue University in 1971 and served as a Postdoctoral Fellow from 1971 to 1973 at the Smithsonian Institution's Radiation Biology Laboratory. Dr. Honeycutt worked as a Senior Chemist at Rohm and Haas Company from 1973 to 1976 and as a Senior Metabolism Chemist at Ciba Geigy from 1976 to 1989. Currently, he is President of the Hazard Evaluation and Regulatory Affairs Company, Inc., which he founded in 1990, and is an analytical biochemist and field research specialist/consultant engaged in exposure assessment of pesticides to humans and the environment.

Dr. Honeycutt is a former chair and council representative for the Central North Carolina Section of the American Chemical Society. He is also a former chair of the Division of Agrochemicals for the American Chemical Society and is currently the Nominations Committee Chair for the Division of Agrochemicals. He served on the ACS Committee on Environmental Improvement for 9 years and is a consultant for this ACS committee. He is also a member of the International Commission of Occupational Health.

Dr. Honeycutt has published widely in the field of pesticide chemistry and has edited several books on this discipline, including: *Journal of Clinical Toxicology — A Special Symposium on Pesticide Metabolites: Analytical and Toxicological Significance*, 19(6/7), i–v, 535–806, 1982–1983, edited by J. Chambers and R.C. Honeycutt; *Dermal Exposure Related to Pesticide Use — Discussion of Risk Assessment*, by R.C. Honeycutt, G. Zweig, and N. Ragsdale, ACS Symposium Series 273, American Chemical Society, 1985; *Evaluation of Pesticides in Groundwater*, by W. Garner, R.C. Honeycutt, and H. Nigg, ACS Symposium Series 315, American Chemical Society, 1986; *Biotechnology in Agricultural Chemistry*, by H. LeBaron, R. Mumma, R.C. Honeycutt, and J. Duesing, ACS Symposium Series 334, American Chemical Society, 1987; *Regulation of Pesticides: Science, Law, and the Media*, edited by R.C. Honeycutt, Government Institutes, Inc., 1988; *Biological Monitoring for Pesticide Exposure Measurement, Estimation and Risk Reduction*, edited by R. Wang, R.C. Honeycutt, C. Franklin, and J. Reinert, ACS Symposium Series 382, American Chemical Society, 1989; *Mechanisms of Pesticide Movement into Groundwater*, edited by R.C. Honeycutt and D.J. Schabacker, Lewis Publishers, Boca Raton, FL, 1994.

Edgar W. Day, Jr., Ph.D., is a native of southern Indiana. He received his B.S. in Chemistry from the University of Notre Dame in 1958 and his Ph.D. in Analytical Chemistry from Iowa State University in 1963. Dr. Day joined the Research Laboratories of Eli Lilly and Company in 1963, working in the fields of pesticide residue chemistry, environmental fate, worker exposure, and risk assessment. In 1989, he joined DowElanco, a joint venture of Lilly and Dow Chemical Company, where he continued his work in field exposure studies and human risk assessment until his retirement in mid-1997.

Dr. Day was a founding member of the Pesticide Handlers Exposure Database (PHED) Task Force, and was instrumental in the development of the database, which is now widely used in the field of pesticide worker exposure. In 1994, he was a key player in the formation of the Agricultural Worker Re-entry Task Force (ARTF) and the Outdoor Residential Exposure Task Force (ORETF), and he served as chairman of the ARTF Technical Committee from its inception in 1994 until his retirement from DowElanco. Dr. Day also served as his company's representative on the Residue Technical Committee of the International Life Sciences Institute (ILSI) from 1989 to 1996. He continues to serve as a consultant to the agrochemical industry.

Contributors

S.C. Artz
Artz Analytical Consulting Services
Durham, North Carolina

G.F. Backhaus
Biologische Bundesanstalt für Land
  und Forstwirtschaft
Braunschweig, Germany

M.J. Bartels
The Dow Chemical Company
H&ES Analytical Chemistry
  Laboratory
Midland, Michigan

D.H. Brouwer
TNO Nutritional and Food Research
Department of Occupational
  Toxicology
Zeist, The Netherlands

B. Chen
DowElanco
Indianapolis, Indiana

W.L. Chen
DowElanco
Indianapolis, Indiana

J.R. Clark
Grayson Research, Ltd.
Durham, North Carolina

C. Colosio
International Centre for Pesticide
  Safety
Milan, Italy

E.W. Day, Jr.
DowElanco
Indianapolis, Indiana

M. DeGeare
H.E.R.A.C., Inc.
Greensboro, North Carolina

M. de Haan
TNO Nutrition and Food Research
Department of Occupational
  Toxicology
Zeist, The Netherlands

S.A.F. De Vreede
TNO Nutrition and Food Research
Department of Occupational
  Toxicology
Zeist, The Netherlands

T.M. Dinoff
Department of Entomology
Personal Chemical Exposure
  Program
University of California
Riverside, California

A. Fait
International Centre for Pesticide
  Safety
Milan, Italy

A. Ferioli
International Centre for Pesticide
  Safety
Milan, Italy

D.F. Hill
Quality Associates, Inc.
Ellicott City, Maryland

E. Hoernicke
Biologische Bundesanstalt für Land
  und Forstwirtschaft
Braunschweig, Germany

M. Honeycutt
H.E.R.A.C., Inc.
Greensboro, North Carolina

R.C. Honeycutt
H.E.R.A.C., Inc.
Greensboro, North Carolina

B. Houtman
DowElanco
Indianapolis, Indiana

R.S. Kludas
Grayson Research, Ltd.
Creedmoor, North Carolina

B. Krebs
Deputy Head, European Regulatory
  Affairs
Insecticides PGRs
Aventis CropScience GmbH
Industriepark Höchst
Frankfurt, Germany

R.I. Krieger
Department of Entomology
Personal Chemical Exposure
  Program
University of California
Riverside, California

W. Maasfeld
Bayer AG
Business Group Crop Protection
Leverkusen, Germany

M. Maroni
International Centre for Pesticide
  Safety
Milan, Italy

W. J.A. Meuling
TNO Nutrition and Food Research
Department of Occupational
  Toxicology
Zeist, The Netherlands

P. Murphy
The Dow Chemical Company
H&ES Analytical Chemistry
  Laboratory
Midland, Michigan

R.J. Nolan
DowElanco
Indianapolis, Indiana

H.-G. Nolting
Biologische Bundesanstalt für Land
  und Forstwirtschaft
Braunschweig, Germany

A. Panzacchi
International Centre for Pesticide
  Safety
Milan, Italy

J.R. Purdy
Ciba-Geigy
Mississauga, Ontario
Canada

J.H. Ross
Department of Pesticide Regulation
California Environmental
  Protection Agency
Sacramento, California

K.D. Schnelle
DowElanco
Indianapolis, Indiana

J. Schrader
Bayer AG
Business Group Crop Protection
Leverkusen, Germany

B.A. Shurdut
DowElanco
Indianapolis, Indiana

P. Swidersky
Quality Associates, Inc.
Ellicott City, Maryland

J.R. Vaccaro
The Dow Chemical Company
H&ES Analytical Chemistry
  Laboratory
Midland, Michigan

J.J. van Hemmen
TNO Nutrition and Food Research
Department of Occupational
  Toxicology
Zeist, The Netherlands

D. Westphal
Bundesinstitut für gesundheitlichen
  Verbraucherschutz and
  Veterinärmedizin
Berlin, Germany

R. Wolf
Residues and Food Safety Specialist
European Regulatory Affairs
Aventis CropScience GmbH
Industriepark Höchst
Frankfurt, Germany

Contents

chapter one

Biological monitoring methods for pesticide exposure evaluation

M. Maroni, C. Colosio, A. Fait, A. Ferioli, and A. Panzacchi

Contents

Summary

Information on exposure levels is fundamental for the assessment and management of health risks related to occupational and environmental exposure to pesticides. Biological monitoring is a primary tool for exposure evaluation,

1

particularly in field users. The authors have reviewed the literature on methods for biological monitoring of pesticide exposure that has been published over the period from 1980 to 1996; summary results are presented for organophosphorus compounds, carbamates, dithiocarbamates, synthetic pyrethroids, organochlorine pesticides, phenoxyacetate herbicides, coumarin rodenticides, atrazine, and pentachlorophenol. Only a minority of the biomarkers addressed in the literature can be used for practical biological monitoring of pesticide exposure. Constraints and problems are discussed in the text, together with needs for further research.

Introduction

Pesticides are chemicals designed to interfere with biological systems and are deliberately spread into the environment. The majority of these compounds are not completely specific for the target organisms, thus they may endanger other species, including humans. Pesticides are used worldwide in the fields of agriculture, industry, and public health and are also used for domestic applications. As a consequence, a great part of the population may be exposed through both the occupational and general environment. In spite of this extensive use, knowledge on the health risks associated with prolonged exposure is rather poor, and great uncertainties still exist. Epidemiological observations in humans have so far produced little conclusive information, primarily because of a weakness in exposure assessment; therefore, information on the type and levels of exposure is fundamental to understanding and characterizing the risks to human health.

Exposure assessment can be carried out through the measurement of environmental concentrations (environmental monitoring), as well as by determination of the chemical or its metabolites in body tissues (biological monitoring). Besides indices of internal dose, biological monitoring also includes measurements of early effects attributable to interactions between chemical agents and the human body. Biological monitoring has the advantage over environmental monitoring of being able to determine the dose actually absorbed through any possible routes; differences in absorption can be taken into account, whether they are due to biological variability or use of protective equipment. Biological monitoring can also consider those cases for which a combination of occupational and non-occupational exposure has occurred.

A few reference documents have been published on biological monitoring of pesticides. For this reason, the Office of Occupational Health of the World Health Organization has given the International Centre for Pesticide Safety (ICPS) the mandate to prepare a monograph specifically addressed to reviewing methods for biological monitoring of pesticide exposure. This chapter has been abstracted from the main review, which is based on more than 300 studies published over the period 1980–1996. For the most representative chemical classes, the available biological exposure indices are reported. Indices of both internal dose and, when available, early effects are discussed. The reported tests have been used to monitor exposure of pesticide applicators in agriculture and public health, manufacturing and formulating workers, subjects poisoned after accidental exposure or attempted suicide, and volunteers involved in pharmacokinetic studies, as well as subgroups of the general population exposed to environmentally persistent pesticides.

Organophosphorus insecticides

Organophosphorus (OP) insecticides cause toxic effects to humans through the inhibition of acetylcholinesterase (ACHE) in the nervous system; this enzyme instantly performs the hydrolytic cleavage of acetylcholine, the chemical mediator responsible for transmission of nerve action potentials. When phosphorylated by OP insecticides, ACHE is no longer able to break down acetylcholine into choline and acetic acid. The resulting accumulation of acetylcholine at different sites in the body is responsible for the typical signs and symptoms of acute OP poisoning (cholinergic overstimulation syndrome) (Dillon and Ho, 1987).

There are different types of cholinesterases in the human body, and they differ in their location in tissues, substrate affinity, and physiological function. The main ones are ACHE, present in nervous tissue and red blood cells (RBC-ACHE), and plasma cholinesterases (PCHE), present in glial cells, plasma, and liver. The physiological functions of RBC-ACHE and PCHE, if any, are unknown.

Organophosphate insecticides also inhibit RBC-ACHE and PCHE. Inhibition of ACHE in erythrocytes is assumed to mirror inhibition of ACHE in the nervous system, which is the receptor of the toxic action, to some extent. Therefore, measurements of RBC-ACHE and PCHE are used for biological monitoring of exposure to OP insecticides (Maroni, 1986). Inhibitions of RBC-ACHE and PCHE activities are correlated with intensity and duration of exposure, although at different levels for each OP compound. Blood ACHE, being the same molecular target as that responsible for acute toxicity in the nervous system, is a true indicator of effect, while PCHE can only be used as an indicator of exposure.

Levels of ACHE and PCHE vary in healthy people because of genetic differences or under specific physio-pathological conditions; inter-individual coefficients of variation of cholinesterase activity have been determined to be about 15 to 25% for PCHE and 10 to 18% for RBC-ACHE. Corresponding figures for intra-individual variations are 6% and 3 to 7%, respectively (Dillon and Ho, 1987).

Because of the physiological variations of RBC-ACHE and PCHE levels, the sensitivity of these tests to detect low-level inhibitions can be increased by comparison with individual pre-exposure values, adopted as a reference. The World Health Organization (WHO, 1982) recommends calculating individual pre-exposure values as the average of three samples; cholinesterase activities after exposure should be expressed as percentage change with respect to individual baseline. Where pre-exposure levels are unknown, RBC-ACHE and PCHE activities should be compared with the activities in a reference population. In this last case, identification of cholinesterase inhibition may be difficult, given the physiological variations of cholinesterase activities.

Intervention measures have been proposed, based on the relationships between ACHE inhibition levels and biological effects (ICOH, 1986; Zielhuis, 1972). An ACHE decrease of 30% or less from the baseline (or 50% or less from the average reference level) requires medical surveillance and examination of working conditions. A reduction of more than 30% from baseline (or 50% from the reference level) requires temporary removal from exposure and careful evaluation of the working conditions.

A fairly good correlation exists between RBC-ACHE inhibition and acute cholinergic effects of severe organophosphate poisoning (Table 1), while such an association is not necessarily present for PCHE. After prolonged low-level exposure, clinical signs of intoxication may appear only at inhibition levels of 85 to 90% of baseline, as opposed to the 60 to 70% inhibition levels usually observed after a single high-level exposure (Jeyaratnam and Maroni, 1994).

Even though all OP insecticides have a common mechanism of action, differences occur among individual compounds. OP insecticides can be grouped into *direct* and *indirect* ACHE inhibitors. Direct inhibitors are effective without any metabolic modification, while indirect inhibitors require biotransformation to be effective. Moreover, some OP pesticides inhibit ACHE more than PCHE, while others do the opposite. For example, malathion, diazinon, and dichlorvos are earlier inhibitors of PCHE than of ACHE. In these cases, PCHE is a more sensitive indicator of exposure, even though it is not correlated with symptoms or signs of toxicity.

Some OP compounds induce delayed neurotoxic effects ("delayed neuropathy") after acute poisoning. This delayed neurotoxic action is independent of cholinesterase inhibition but related to phosphorylation of a specific esterasic enzyme in the nervous tissue, called "neurotoxic esterase" or "neuropathy target esterase" (NTE) (Johnson, 1982). NTE is present in the nervous tissue, liver lymphocytes, platelets, and other tissues, but its physiological function is unknown. There is a rather large inter-individual variation of lymphocyte and platelet NTE activity (Table 2).

No data are available on platelet NTE activity in exposed subjects, and little data on lymphocyte NTE activity. In one reported case of suicidal poisoning with chlorpyrifos, inhibition of lymphocyte NTE was correlated with the enzyme inhibition in peripheral nerves (Osterloh et al., 1983). In another case of attempted suicide with the same compound, inhibition of NTE in peripheral lymphocytes was associated with the development of delayed neuropathy (Lotti, 1986). However, the threshold of NTE inhibition required for delayed neuropathy remains undetermined (Lotti, 1987). Observations in occupationally exposed subjects are limited in number, and more research is needed to investigate the applicability of NTE as a biomarker of exposure to OP pesticides.

Table 1 Severity and Prognosis of Acute OP Intoxication
at Different Levels of ACHE Inhibition

% ACHE inhibition	Level of poisoning	Clinical symptoms	Prognosis
50–60	Mild	Weakness, headache, dizziness, nausea, salivation, lacrimation, miosis, moderate bronchial spasm	Convalescence in 1–3 days
60–90	Moderate	Abrupt weakness, visual disturbance, excess of salivation, sweating, vomiting, diarrhea, bradycardia, hypertonia, tremors of hands and head, disturbed gait, miosis, chest pain, cyanosis of the mucous membranes	Convalescence in 1–2 weeks
90–100	Severe	Abrupt tremor, generalized convulsions, psychic disturbance, intensive cyanosis, lung edema, coma	Respiratory or cardiac failure, death

Table 2 Reference Values of Lymphocyte and
Platelet Neuropathy Target Esterase Activity

Location	No. of subjects	Lymphocytes (nmol min^{-1} mg $prot^{-1}$) (mean ± SD)	VC	Platelets (nmol min^{-1} mg $prot^{-1}$) (mean ± SD)	VC	Ref.
Baltimore, MD, 1986	68	13.2 ± 2.4	18	8.4 ± 1.5	18	Maroni and Bleecker (1986)
Valtellina, Italy, 1985	35	11.9 ± 3.0	25	8.5 ± 1.9	22	Maroni et al. (1985)
Padova, Italy, 1985	108	11.5 ± 2.5	22	—	—	Bertoncin et al. (1985)

Metabolism of most OP pesticides yields alkylphosphates or alkyl-(di)-thiophosphates as a result of the hydrolysis of the P-X bond in the OP molecule. Alkylphosphates detectable in urine, and the respective parent compound from which they can originate, are listed in Table 3. Because these metabolites are common to several different OP pesticides, this method is not compound specific and is usable only to assess exposure to all the parent compounds that generate these derivatives.

Besides alkylphosphates, OP metabolism gives rise to the production of other metabolites that can be used as exposure markers (Table 4). Unchanged OP compounds in blood or urine can also be measured to confirm exposure (Table 4), but this method is of limited use for routine biological monitoring of occupational exposure, as OP compounds are rapidly excreted in urine. Moreover, most OP pesticides are unstable, and, with a few exceptions, they are not detectable in biological specimens after a few hours. So far, the measurement of unchanged compounds in biological fluids has been performed primarily for research purposes and has limited practical applicability.

Table 3 Alkylphosphates Detectable in Urine as Metabolites
of Some Organophosphorus Pesticides

Metabolite	Abbreviation	Principal parent compounds
Monomethylphosphate	MMP	Malathion
Diethyldithiophosphate	DEDTP	Disulfoton, phorate
Diethylphorphorothiolate	DEPTh	Disulfoton, phorate
Diethylphosphate	DEP	Parathion, demethon, diazinon, dichlofenthion
Diethylthiophosphate	DETP	Diazinon, demethon, parathion
Dimethyldithiophosphate	DMDTP	Malathion, dimethoate, azinphos-methyl
Dimethylphosphate	DMP	Dichlorvos, trichlorfon, mevinphos, malathion, dimethoate, fenchlorphos, monocrotophotos
Dimethylthiophosphate	DMTP	Fenitrothion, fenchlorphos, malathion, dimethoate, azinphos-methyl
Phenylphosphoric acid	PPA	Leptophos, EPN

Carbamate pesticides

Carbamates are used as insecticides, nematocides, fungicides, and herbicides; the toxicity of carbamate insecticides is similar to that of OP compounds and is based on the inhibition of ACHE. Also, carbamate metabolites may inhibit ACHE but are usually weaker inhibitors than the unchanged compound. Cholinesterase inhibition caused by carbamates is labile, of short duration, and rapidly reversible; in fact, the half-life of the inhibited enzymes ranges between some minutes and 2 to 3 hours for RBC-ACHE and is on the order of some minutes for PCHE. Accumulation of cholinesterase activity on repeated exposures, as observed with OP compounds, does not occur with

Table 4 Other Biomarkers Used in Biological Monitoring
of Human Exposure to Organophosphates

Compound	Biological indicators	Sample[a]
Acephate	Acephate	U
Azinphos-ethyl	Azinphos-ethyl	B
Chlorfenvinphos	Desethyl-chlor	U
Bromophos	Bromophos	B
Chlorpyrifos	3,5,6-Trichloro-pyridinol (TCP)	B/U
Dichlofenthion	Dichlofenthion	B
	2,4-Dichloropropane	U
Dimethoate	Dimethoate	B
Fenitrothion	3-Methyl-4-nitrophenol (3-4-NP)	U
	Fenitrothion	B
Formothion	Dimethoate	B
Glyphosate	Glyphosate	U
	Aminomethyl-phosphonic acid	U
Parathion	*p*-Nitrophenol	U
	Parathion	B
	Paraxon	B
Methyl-parathion	*p*-Nitrophenol	U
Trichlorfon	Trichlorfon	B/U

[a] B = blood; U = urine.

carbamates. Scarce information is available for the ACHE inhibition levels associated with symptoms; cholinergic symptoms have been usually observed in carbamate-exposed workers with blood ACHE activity lower than 70% of individual baseline (Baron, 1991; WHO, 1986, 1994).

Occupational exposure to carbamate insecticides may be monitored by measuring RBC-ACHE and/or PCHE. However, given the low cholinesterase inhibition levels and the short time duration of this effect, ACHE inhibition can generally be used as a biomarker of exposure only when exposure levels are high. Three sequential samples are recommended to establish an individual baseline before exposure. In exposed workers, blood sampling and analysis should be carried out soon after the end of exposure (WHO, 1986).

Unchanged compounds or metabolites in blood and urine can be used to monitor human exposure to some carbamates. Table 5 shows some biological indices of internal dose used to monitor carbamate exposure. Urine carbamate metabolites may provide a good estimate of the internal dose; because the half-life of most compounds is very short, samples collected soon after the end of the exposure are preferable for analysis (WHO, 1986).

Dithiocarbamates

Dithiocarbamate (DTC) pesticides are primarily used in agriculture as fungicides, insecticides, and herbicides. Additional uses are as biocides for

Table 5 Biomarkers Used in Biological Monitoring
of Human Exposure to Carbamates

Compound	Biological indicators	Sample[a]
Carbamates	ACHE, PCHE	B
Aldicarb	Aldicarb-sulfone	U
Carbaryl	1-Naphtol	U
	Carbaryl	B
Methomyl	Methomyl	B
Propoxur	2-Isopropoxyphenol (2-IPP)	U
	Propoxur	B

[a] B = blood; U = urine.

industrial applications and in household products. Some DTC are used for vector control in public health.

Dithiocarbamates have a complex metabolism. A metabolite common to all of them is carbon disulfide (CS_2), which is later metabolized to 2-thiothiazolidine-4-carboxylic acid (TTCA) (WHO, 1988). Carbon disulfide is primarily excreted through exhaled air and, to a lesser extent, through urine. Measurement of urine levels of CS_2 has been suggested to monitor high-level exposure to these compounds (Table 6). Urine samples should be collected in the morning following cessation of exposure (Liesivuori and Savolainen, 1994). Determination of CS_2 prior to exposure has been recommended to establish individual baselines, as CS_2 is commonly also found in urine of non-exposed subjects (Brugnone et al., 1992).

The lack of specificity of this biomarker and the complexity of analytical procedures for CS_2 determination represent major limitations to the practical use of this metabolite to monitor occupational exposure. Moreover, only few data are available to confirm the validity of CS_2 as a biomarker of DTC exposure.

Ethylenethiourea (ETU) is one of the metabolic products of ethylene-*bis*-dithiocarbamates (EBDTC) in mammals, plants, and lower organisms (WHO, 1988). It may also be present as an impurity in EBDTC technical formulations. EBDTC residues present in food may be partially transformed into ETU during food preparation.

A few studies have been carried out to investigate ETU concentrations in urine of workers exposed to EBDTC (Kurttio et al., 1988, 1990) (Table 6). According to Kurttio, ETU determination should be performed in 24-hr urine samples collected from the end of exposure. The results of these studies show that the analytical method necessary to measure ETU in urine is time consuming and rather complicated. The use of ETU as a biomarker of EBDTC exposure is a promising tool for biological monitoring of exposed workers, but, in light of the analytical complexity, this index is barely usable for routine monitoring.

Table 6 Biomarkers Used in Biological Monitoring of
Human Exposure to Some Groups of Pesticides

Compound	Biological indicators	Sample
Dithiocarbamates	Carbon disulfide CS$_2$ (?)	U
	2-Thiothiazolidine-4-carboxylic acid (2-4-TTCA) (?)	U
Methiram		
Thiram	Xanthurenic acid	U
Ziram		
EBDTC (ethylene-bis-dithiocarbamates)	Ethylenthiourea (ETU)	U
	ETU-hemoglobin adducts (?)	B
	Carbon disulfide CS$_2$ (?)	U
	2-Thiothiazolidine-4-carboxylic acid (2-4-TTCA) (?)	U
Mancozeb	Mn (?)	U
Maneb	Mn (?)	U
Propineb		
Zineb		
Phenoxyacetates		
2,4-Dichlorophenoxyacetic acid (2,4-D)	2,4-D	B/U
2,4,5-Trichlorophenoxyacetic acid (2,4,5-T)	2,4,5-T	B/U
Dichlorprop	Dichlorprop	U
4-Chloro-2-methylphenoxyacetic acid (MCPA)	MCPA	B/U
Picloram	Picloram	U
Silvex	Silvex	B/U
Quaternary ammonium compounds		
Diquat	Diquat	B/U
Paraquat	Paraquat	B/U
Coumarins	Prothrombin, prothrombin time	B
Brodifacoum	Brodifacoum	B
Bromadiolone	Bromadiolone	B
Chlorophacinone	Chlorophacinone	B
Difenacoum	Difenacoum	B
Others		
Chlorotriazines	Chlorotriazines	U
	Dealkylated metabolites	U
	Mercapturic acid derivatives	U
Pentachlorophenol (PCP)	PCP	B/U

[a] B = blood; U = urine.

Some authors have recently shown that ETU reacts with hemoglobin and leads to the development of hemoglobin-ETU adducts in subjects occupationally exposed to mancozeb, an ethylene-*bis*-dithiocarbamate pesticide (Colosio et al., 1996; Pastorelli et al., 1995); this finding suggests the possibility of using hemoglobin-ETU adducts for biological monitoring of exposure to EBDTC.

Dithiocarbamates are chemically characterized by the presence of metals in the molecule (iron, manganese, zinc, etc.); therefore, the measurement of these metals in urine has been proposed as an alternative approach to monitor exposure. For instance, increased urinary excretion of manganese has been reported in workers exposed to mancozeb (Canossa et al., 1993). Available data are at present insufficient to confirm the possibility of using metals as biomarkers of human exposure to DTC.

Phenoxyacids

Phenoxyacids (PA) have been widely used as herbicides in agriculture, forestry, and, to a lesser extent, garden activities. The principal products are represented by 2,4-D (2,4-dichlorophenoxyacetic acid); 2,4,5-T (2,4,5-trichlorophenoxyacetic acid); and MCPA (4-chloro, 2-methylphenoxyacetic acid) (Stevens and Sumner, 1991). 2,4,5-T has been banned in many countries for a long time because of contamination of the commercial formulations of 2,4,5-T by 2,3,7,8-tetrachlorodibenzodioxin. At present, dioxin contamination of these formulations has been reduced to very low concentrations.

Phenoxyacid herbicides are poorly metabolized by mammalian species; most of the absorbed PA herbicides are excreted in urine unmodified or as conjugate derivatives. Biological indicators to monitor exposure to PA are shown in Table 6. Urine is the recommended medium for the determination of unchanged compounds. Plasma concentrations can be measured to assess acute poisoning cases or, at autopsy, to confirm absorption of these compounds. In fatal poisonings, blood concentrations higher than 100 mg/L have been detected. Knowledge about the relationship between exposure levels and internal doses of PA herbicides is confined to a few compounds, as these herbicides are characterized by a limited use in agriculture. The urinary excretion rate of PA herbicides varies according to their route of absorption. After ingestion, most of the absorbed dose is excreted within 24 hr, while, after dermal exposure, urine excretion is delayed up to 36 to 48 hr (Kolmodin-Hedmann, 1994). More than 80% of the absorbed dose is generally excreted within 5 days after exposure.

Under field exposure conditions, it is recommended to measure PA herbicides in 24-hr urine samples collected starting at the end of the workshift. Spot samples collected at the end of exposure or the following morning can be used when a 24-hr urine collection is impractical. In this case, the concentration of the compounds should be normalized to creatinine concentration or adjusted for specific gravity.

Quaternary ammonium compounds

Diquat and paraquat are quaternary ammonium compounds largely used as contact herbicides and crop desiccants. When systemic absorption occurs, paraquat and diquat are rapidly distributed into the body. Paraquat primarily accumulates in the lungs and kidneys, while the highest diquat concentrations have been found in the gastrointestinal tract, liver, and kidneys (WHO, 1984). Urine is the principal route of excretion for both diquat and paraquat, which are primarily eliminated as unmodified compounds. Occupationally exposed workers can be monitored by measuring paraquat and diquat concentrations in urine samples (Table 6). Blood concentrations are useful to monitor acute poisoning cases.

Coumarin rodenticides

Coumarin rodenticides, also known as vitamin K-antagonists, act as anticoagulants in various mammalian species, including rats and mice. These compounds can be grouped into first-generation (warfarin) and second-generation (brodifacoum) vitamin K-antagonists, the latter being characterized by biological half-lives on the order of 100 to 200 days. Coumarin derivatives act by inhibiting the enzymes of the vitamin K cycle in the liver (vitamin K epoxide [KO]-reductase and vitamin K-reductase), blocking the recycling of vitamin K and thereby depleting the supply of vitamin KH_2, the active form of vitamin K. Administration of coumarin derivatives results in a decrease in the rate of synthesis of the vitamin K-dependent clotting factors II, VII, IX, and X (Pelfrene, 1991).

Biological monitoring of exposure to coumarin derivatives can be performed by determination of the unchanged compound and/or its metabolites in blood and urine (Table 6). Analytical complexity and the lack of knowledge about the dose-effect relationship in exposed subjects are the primary limitations of this method.

Alternatively, biological monitoring may be carried out by determining the effects of exposure on the coagulation system. To this purpose, two kinds of tests are available. One test is based on measurement of depressed coagulation (i.e., circulating clotting factors decrease below 20% of normal); the other type of test monitors early effects of exposure (i.e., circulating clotting factors decrease at 70 to 80% of normal plasma levels). Prothrombin time is most sensitive to monitor clinically depressed coagulation and may be used in the occupational setting to monitor the severity of an existing overexposure. However, this test is not sensitive enough for detection of early effects of exposure. To that purpose, determination of the prothrombin concentration (factor II) in plasma is recommended (Van Sittert and Tuinman, 1994).

In the future, this test may be replaced by the proteins induced by vitamin K antagonists absence (PIVKA) concentrations in serum. PIVKAs are coagulation factor precursors, normally not detectable in blood but released into the blood stream in case of blockage of the vitamin K cycle by coumarins. So far, this test has only been used for research purposes (Greef et al., 1987).

Synthetic pyrethroids

Synthetic pyrethroids are esters of specific acids (chrysantemic acid, halo-substituted chrysantemic acid, 2-[4-chlorophenyl]-3-methylbutyric acid) and alcohols (allethrone, 3-phenoxybenzyl alcohol) (Ray, 1991). They represent a group of insecticides largely used in agriculture and public health because of their relatively low toxicity to humans and mammalian species and their short environmental persistence.

Pyrethroid insecticides are rapidly metabolized to their inactive acids and alcohol components, which are excreted primarily in urine. A small portion of the absorbed compounds is excreted unchanged. Occupational exposure to pyrethroid insecticides can be assessed by measuring intact compounds or their metabolites in urine. Biological indicators of internal dose in exposed subjects are reported in Table 7. Due to their rapid metabolism, determination of blood concentrations can only be used to reveal recent high-level exposures.

The relationship between exposure and internal dose is known only for a few pyrethroids. Human volunteer studies have shown that, after a single oral administration, pyrethroids and the respective metabolites are excreted in urine within 24 hr and do not accumulate in the body. In field workers exposed to cypermetrin through the dermal route, urine excretion of the intact compound and its metabolites peaked 36 hr after exposure had ceased (WHO, 1989).

Organochlorine pesticides

Organochlorine (OC) compounds represent a class of pesticides primarily used as insecticides (Smith, 1991). They were widely used in the 1950s and 1960s, but the use of OC compounds has been discontinued in many countries because of their high environmental and biological persistence. According to their chemical structure, OC compounds can be divided into the following sub-groups: (1) benzene hexachloride isomers, such as lindane; (2) cyclodienes, such as aldrin, dieldrin, endrin, chlordane, heptachlor, and endosulfan; (3) DDT and analogs, such as methoxychlor, dicofol, and chlorobenzilate.

After absorption, OC compounds are rapidly distributed and accumulate in high-fat-content tissues; the degree of accumulation is inversely related to the rate of biotransformation into water-soluble metabolites. The biological half-lives range from a minimum value of 24 hr for lindane, to 1 year for dieldrin, to 3–4 years for DDT (Tordoir and Van Sittert, 1994).

As a consequence of their environmental persistence, most OC pesticides have become ubiquitous pollutants; therefore, these compounds are detectable in trace amounts in biological tissues of higher animal species, including

Table 7 Biomarkers Used in Biological Monitoring
of Human Exposure to Pyrethroids

Compound	Biological indicators (urine)
Cypermethrin	Cl_2CA,[a] 4-HPBA[b]
Deltamethrin	Deltamethrin
	Br_2CA[c]
Fenvalerate	Fenvalerate
Permethrin	Cl_2CA

[a] *cis-* and *trans-*3-(2,2'-dichloro-vinyl)-2,2'-dimethyl-
cyclopropane carboxylic acid.
[b] 3-(4'-hydroxyphenoxy) benzoic acid.
[c] Dibromovinyl-dimethyl-cyclopropane carboxylic acid.

humans. Detectable amounts of banned compounds are still found in fluids and tissues of the general population in countries where they were used in the past.

Biological indicators of the internal dose of OC pesticides are listed in Table 8. Biological monitoring of OC exposure can be carried out through determination of the presence of OC compounds or their metabolites in blood and urine. In general, measurements of highly lipophilic organochlorines (such as aldrin and DDT) and specific metabolites in blood and urine are valuable methods for monitoring acute poisoning cases or occupational high-level exposures when a steady-state condition is reached. Other OC compounds, such as dichloropropene, that are rapidly metabolized into water-soluble compounds can be detected in urine of exposed workers shortly after the beginning of exposure.

In light of their high lipophilic properties, many OC compounds and some of their metabolites are detectable in adipose tissue; however, this method cannot be applied to monitor occupationally exposed subjects on a routine basis because of the invasive nature of the sampling procedures. Because intact OC compounds and their metabolites are commonly found in blood and urine of the general population, comparison with appropriate reference groups or with individual pre-exposure values is recommended when surveying occupationally exposed workers.

Some OC pesticides can induce the hepatic microsomal enzyme system (Kay, 1970). Tests measuring functions related to these enzymes, such as f.i. D-glucaric acid and 6-b-hydroxicortisol excretion in urine, can be applied to monitor occupational OC exposure.

Because hexachlorobenzene can interfere with porphyrin metabolism and produce chemical porphyria, the evaluation of the urinary porphyrin pattern has been proposed as an early biomarker of effect (San Martin et al., 1977; Schmid, 1960, 1966; Wray et al., 1962). However, apart from an outbreak of chemical porphyria which occurred in Turkey in the 1950s following ingestion of hexachlorobenzene-contaminated food, cases of chemical porphyria among

Table 8 Biomarkers Used in Biological Monitoring of
Human Exposure to Organochlorines

Compound	Biological indicators	Sample
Aldrin and dieldrin	Dieldrin	B
Chlordecone	Chlordecone	B
Chlordane	Chlordane and metabolites	B
Chlorobenzilate	p-p'-Dichlorobenzophenone (DBP)	B
DDT	DDT, DDE	B/U
1,3-Dichloropropene	DCP-MA[a]	U
	Thioethers	U
Endosulfan	Endosulfan	B
Endrin	Endrin	B
	Anti-12-hydroxy-endrin	U
Heptachlor	Heptachlor epoxide	B
Hexachlorobenzene	Porphyrin pattern	U
Lindane	Lindane	B
Toxaphene	Toxaphene	B

[a] N-acetyl-S-(*trans*-3-chloroprop-2-enyl)-cysteine and N-acetyl-S-(*cis*-3-chloroprop-2-enyl)-cysteine mercapturic acid.

occupationally exposed workers have not been reported. Hexachlorobenzene is probably able to induce chemical porphyria only at very high doses that are the result of acute overexposure; thus, measurement of urinary porphyrins is of limited value to monitor occupationally exposed workers.

Chlorotriazines

Atrazine, used as a selective pre- and post-emergence herbicide to control annual weeds in several crops, is the most representative compound of this group. It is also used as a non-selective herbicide in non-crop areas. After absorption, the compound is metabolized to dealkylated and deisopropylated derivatives. The unchanged compound and its metabolites are excreted in urine, where they can be detected by chromatography or enzyme-linked immunosorbent assay (Lucas et al., 1993). A mercapturic acid conjugate of atrazine has also been found in urine samples of workers spraying this herbicide (Lucas et al., 1993) (Table 6).

Excretion of free atrazine in urine is consistent with the pattern of exposure, with maximal excretion rates at the end of the workshift and a rapid decrease after cessation of exposure. This pattern suggests that atrazine does not accumulate in the body (Catenacci et al., 1990). Atrazine metabolism gives rise to bi-dealkylated (80%), deisopropylated (10%), and deethylated (8%) metabolites, which are eliminated in urine over a period slightly longer than 24 hr. Fifty percent of the absorbed dose is excreted within 8 hr (Cate-nacci et al., 1993). For analysis, 24-hr urine samples collected from the beginning of exposure are suitable.

Other chlorotriazines (simazine, propazine, terbuthylazine) follow the same biotransformation pathway of atrazine; therefore, urinary excretion of bi-dealkylated, deisopropylated, and deethylated metabolites is not compound specific. When simultaneous exposure to different chlorotriazines occurs, the unmodified compound measured in urine, even though it represents a minor portion of the absorbed dose, may be useful for a qualitative confirmation of exposure.

Pentachlorophenol

Pentachlorophenol (PCP) is a halogenated phenolic compound used for termite control, as a wood preservative, and as a general herbicide. The sodium PCP salt has been used as a general disinfectant. After absorption, PCP is conjugated with glucuronic acid. The metabolites and the unchanged compound are primarily excreted in urine. The pharmacokinetics of PCP have been investigated in human volunteers orally administered with single or multiple doses of 0.1 mg/kg bw. After a single oral administration, 86% of the PCP dose was excreted almost completely within 7 days, with 74% unmodified and 12% as a glucuronide conjugate. Results obtained after multiple dosing showed that storage and excretion during exposure would reach a steady state in about 8 days with a maximal PCP plasma concentration of 0.5 ppm (Braun et al., 1979).

Pentachlorophenol concentrations in urine and serum can be used as biomarkers of internal dose (Colosio et al., 1993a). PCP concentrations up to about 30 mg/L were detected in urine samples of exposed workers, while concentrations lower than 0.3 mg/L were detected in the general population. The presence of PCP in biological samples of the general population is attributable to indoor exposure to the compound released from treated materials (furniture, leather, paints, etc.).

Less information is available on PCP concentrations in serum. PCP concentrations lower than 0.5 mg/L have been observed in non-exposed populations. Values ranging from 0.2 to 1.8 mg/L have been measured in occupants of PCP-treated houses, and concentrations from 0.4 to 13 mg/L have been measured in occupationally exposed workers. PCP concentrations higher than 40 mg/L have been observed in fatal poisonings.

A good correlation exists between PCP external exposure levels and biomarkers of internal dose. In woodworkers with prolonged exposure to PCP, significant relationships were observed between PCP concentrations on skin (mean values, 23 to 627 $\mu g/dm^2$) and urinary PCP excretion measured in spot samples collected the morning following exposure (mean values, 28 to 175 $\mu g/g$ creatinine). A significant correlation was also observed between dermal and plasma concentrations (mean values, 129 to 703 $\mu g/L$) (Colosio et al., 1993a,b).

Conclusions

There is a growing need to better characterize the health risk related to occupational and environmental exposure to pesticides. Risk characterization is a basic step in the assessment and management of the health risks related to chemicals (Tordoir and Maroni, 1994). Evaluation of exposure, which may be performed through environmental and biological monitoring, is a fundamental component of risk assessment. Biomarkers are useful tools that may be used in risk assessment to confirm exposure or to quantify it by estimating the internal dose. Besides their use in risk assessment, biomarkers also represent a fundamental tool to improve the effectiveness of medical and epidemiological surveillance.

In theory, a "good" biomarker must be sensitive and specific, and the dose-effect relationship must be known. In the case of pesticides, dose-effect relationships are well known for only a minority of compounds (e.g., organophosphates), and biological exposure limits have been established in few cases. As a consequence, biological exposure monitoring can be carried out on a routine basis only for exposure to a few compounds. In many cases, in spite of high levels of sensitivity and specificity of the biomarkers, the analytical complexity or high laboratory costs represent major limitations for routine application of biological indices in the occupational practice. Besides the above-mentioned constraints for the adoption of biological monitoring as a routine tool, biomarkers have been studied only in experimental systems and still remain to be evaluated in humans for a large number of pesticides currently in use.

Duration and route of exposure can influence the interpretation of biological markers; for instance, in some cases, pesticides that are inhaled remain in systemic circulation longer than those that are ingested. Compounds characterized by a slow process of dermal absorption can give rise to prolonged kinetics if compared to the kinetics following ingestion or inhalation of the same compounds.

In occupational settings, the pattern of exposure to pesticides differs between pesticide applicators and industrial workers; industrial workers are usually exposed to one or a few active ingredients for prolonged periods of time, primarily through the respiratory tract, and the degree of exposure is relatively constant over time. Agricultural workers are typically exposed to numerous compounds for short periods of time, primarily through dermal absorption, and the degree of exposure is extremely variable.

There is a growing need to identify markers usable for biological monitoring of prolonged exposures to low doses of chemicals and exposures fluctuating over time. In addition to the traditional indicators, new generations of biomarkers are currently under study. A number of pesticides can be biotransformed into reactive metabolites able to bind covalently biological macromolecules (DNA, proteins, hemoglobin) with the production of macromolecular adducts. These adducts are promising biomarkers, as they provide integrated information that takes into account the ability of the organism to form reactive metabolites and generate a biological effective dose. Moreover, in light of the stable linkage to proteins with a prolonged biological half-life, these markers also provide an opportunity to integrate exposures that fluctuate over time or those lasting for prolonged periods.

Recent investigations have shown that a number of pesticides, such as dithiocarbamates, pentachlorophenol, and organotin compounds, cause immune changes even at low exposure levels. The observed effects are usually slight and consist of changes in the lymphocyte subsets, in the proliferative response to mitogens, and in production of cytokines. The biological relevance of these findings is still unclear, because the investigated subjects did not show clinical evidence of any immune-mediated disease. Even though these observations are not predictive of disease, the available data suggest the possibility of using immune biomarkers for the biological monitoring of occupational exposure to pesticides.

References

Baron R.L. (1991) Carbamate insecticides, in *Handbook of Pesticide Toxicology*, Vol. 2, W.J. Hayes and E.R. Laws (Eds.), Academic Press, San Diego, CA, pp. 917–1123.

Bertoncin, D., Russolo, A., Caroldi, S., and Lotti, M. (1985) Neuropathy target esterase in human lymphocytes, *Archives on Environmental Health*, 40:139–143.

Braun, W.H., Blau, G.E., and Chenoweth, M.B. (1979) The metabolism/pharmacokinetics of pentachlorophenol in man, and a comparison with the rat and the monkey, *Developments in Toxicology and Environmental Science*, 4:289–296.

Brugnone, F., Maranelli, G., Zotti, S., Zanella, I., De Paris, P., Caroldi, S., and Betta, A. (1992) Blood concentration of carbon disulfide in "normal" people and after antabuse treatment, *British Journal of Industrial Medicine*, 49:658–663.

Canossa, E., Angiuli, G., Garasto, G., Buzzoni, A., and De Rosa, E. (1993) Indicatori di dose in agricoltori esposti a mancozeb, *La Medicina del Lavoro*, 84(1):42–50.

Catenacci, G., Maroni, M., Cottica, D., and Pozzoli, L. (1990) Assessment of human exposure to atrazine through the determination of free atrazine in urine, *Bulletin of Environmental Contamination and Toxicology*, 44:1–7.

Catenacci, G., Barbieri, F., Bersani, M., Ferioli, A., Cottica, D., and Maroni, M. (1993) Biological monitoring of human exposure to atrazine, *Toxicology Letters*, 69:217–222.

Colosio, C., Barbieri, F., Bersani, M., Schlitt, H., and Maroni, M. (1993a) Markers of occupational exposure to pentachlorophenol, *Bulletin of Environmental Contamination and Toxicology*, 51:820–826.

Colosio, C., Maroni, M., Barcellini, W., Meroni, P., Alcini, D., Colombi, A., Cavallo, D., and Foá, V. (1993b) Toxicological and immune findings in workers exposed to pentachlorophenol (PCP). *Archives of Environmental Health*, 48(2):81–88.

Colosio, C., Barcellini, W., Maroni, M. et al. (1996) Immunomodulatory effects of occupational exposure to mancozeb, *Archives of Environmental Health*, 51(6):445–451.

Dillon, H.K. and Ho, M.H. (1987) Biological monitoring of exposure to organophosphorus pesticides, in *Biological Monitoring of Exposure to Chemicals: Organic Compounds*, H.K. Dillon and M.H. Ho (Eds.), Wiley, New York, pp. 227–287.

Gallo, M.A. and Lawryk, N.J. (1991) Organic phosphorous pesticides, in *Handbook of Pesticide Toxicology*, Vol. 2, W.J. Hayes and E.R. Laws (Eds.), Academic Press, San Diego, CA, pp. 917–1123.

Greeff, M.C., Mashile, O., and Macdougall, L.G. (1987) "Superwarfarin" (bromodialone) poisoning in two children resulting in prolonged anticoagulation, *Lancet*, 2:1269.

ICOH (International Commission on Occupational Health) (1986) Biological monitoring of workers manufacturing, formulating, and applying pesticides, in Proceedings of the 7th International Workshop of the Scientific Committee on Pesticides, Szeged, Hungary, April 15–17, E.A.H. van Heemstra-Lequin and N.J. Van Sittert (Eds.), *Toxicology Letters*, 33.

Jeyaratnam, J. and Maroni, M. (1994) Organophosphorus compounds, in *Health Surveillance of Pesticide Workers: A Manual for Occupational Health Professionals*, W. Tordoir, M. Maroni, and F. He (Eds.), Elsevier, Amsterdam, pp. 15–27.

Johnson, M.K. (1982) The target of initiation of delayed neurotoxicity by organophosphorus esters: biochemical studies and toxicological applications, *Reviews in Biochemical Toxicology*, 4:141–212.

Kay, K. (1970) Pesticides and associated health factors in agricultural environments: effects on mixed-function oxiding enzyme metabolism, pulmonary surfactant and immunological reactions, in *Pesticides Symposia*, R. Dichmann (Ed.), Halos & Co., Miami, FL.

Kolmodin-Hedman, B. (1994) Phenoxyacetates, in *Health Surveillance of Pesticide Workers: A Manual for Occupational Health Professionals*, W. Tordoir, M. Maroni, and F. He (Eds.), Elsevier, Amsterdam, pp. 87–91.

Kurttio, P., Vartianen, T., and Savolainen, K. (1988) High-performance liquid chromatographic method for the determination of ethylenthiourea in urine and on filters, *Analytical Chemistry Acta*, 212:297–301.

Kurttio, P., Vartianen T., and Savolainen, K. (1990) Environmental and biological monitoring of exposure to ethylene-*bis*-dithiocarbamate fungicides and ethylenethiourea. *British Journal of Industrial Medicine*, 47:203–206.

Liesivuori, J. and Savolainen, K. (1994) Dithiocarbamates, in *Health Surveillance of Pesticide Workers. A Manual for Occupational Health Professionals*, W. Tordoir, M. Maroni, and F. He (Eds.), Elsevier, Amsterdam, pp. 37–41.

Lotti, M. (1986) Biological monitoring for organophosphate-induced delayed polyneuropathy, *Toxicology Letters*, 33:167–172.

Lotti, M. (1987) Organophosphate-induced delayed polyneuropathy in humans: perspectives for biomonitoring, *Trends in Pharmacological Sciences*, 8:176–177.

Lucas, A.D., Jones, A.D., Goodrow, M.H., Saiz, S.G., Blewett, C., Seiber, J.N., and Hammock, B.D. (1993) Determination of atrazine metabolites in human urine: development of a biomarker of exposure, *Chemical Research in Toxicology*, 6:107–116.

Maroni, M. (1986) Organophosphorus pesticides, in *Biological Indicators for the Assessment of Human Exposure to Industrial Chemicals*, L. Alessio, A. Berlin, M. Boni, and R. Roi (Eds.), Commission of the European Communities (EUR 10704), Brussels, Lusembourg, pp. 51–72.

Maroni, M. and Bleecker, M.L. (1986) Neuropathy target esterase in human lymphocytes and platelets, *Journal of Applied Toxicology*, 6:1–7.

Maroni, M., Colombi, A., Gilioli, R., Bleecker, M.L., Villa, L., and Foà, V. (1985) Reference values for lymphocyte and platelet NTE activity in American and Italian populations, in *Proceedings of the Second International Symposium on Neurobehavioural Methods in Occupational and Environmental Health*, Copenhagen, August 6–9, 1985.

Osterloh, J.D, Lotti, M., and Pond, S.M. (1983) Toxicologic studies in a fatal overdose of 2,4-D, MCPP, and chlorpyrifos, *Journal of Applied Toxicology*, 7:125–129.

Pastorelli, R., Allevi, R., Romagnano, S., Meli, G., Fanelli, R., and Airoldi, R. (1995) Gas chromatography-mass spectrometry determination of ethylenthiourea hemoglobin adducts: a possible indicator of exposure to ethylene-*bis*-dithiocarbamate pesticides, *Archives of Toxicology*, 69(5):306–311.

Pelfrene, A.F. (1991) Synthetic organic rodenticides, in *Handbook of Pesticide Toxicology*, Vol. 3, W.J. Hayes and E.R. Laws (Eds.), Academic Press, San Diego, CA, pp. 1290–1305.

Ray, D.E. (1991) Pesticides derived from plants and other organisms, in *Handbook of Pesticide Toxicology*, Vol. 2, W.J. Hayes and E.R. Laws (Eds.), Academic Press, San Diego, CA, pp. 585–599.

San Martin de Viale, L.C., Rios de Molina, M., Wainstok de Calmanovici, R., and Tomio, J.M. (1977) Porphyrins and porphyrinogen carboxylase in hexachlorobenzene-induced porphyria, *Biochemical Journal*, 168:393–400.

Schmid, R. (1960) Medical intelligence: cutaneous porphyria in Turkey, *New England Journal of Medicine*, 263:397–398.

Schmid, R. (1966) The porphyrias, in *The Metabolic Basis of Inherited Disease*, 2nd ed., J.B. Stanbury, J.B. Wyngaarden and D.S. Frederickson (Eds.), McGraw-Hill, New York, pp. 813–870.

Smith, A.G. (1991) Chlorinated hydrocarbon insecticides, in *Handbook of Pesticide Toxicology*, Vol. 2, W.J. Hayes, and E.R. Laws (Eds.), Academic Press, San Diego, CA, pp. 731–915.

Stevens, J.T. and Sumner, D.D. (1991) Herbicides, in *Handbook of Pesticide Toxicology*, Vol. 3, W.J. Hayes and E.R. Laws (Eds.), Academic Press, San Diego, CA, pp. 1317–1408.

Tordoir, W. and Maroni, M. (1994) Basic concepts in the occupational health management of pesticide workers, in *Health Surveillance of Pesticide Workers: A Manual for Occupational Health Professionals*, W. Tordoir, M. Maroni, and F. He (Eds.), Elsevier, Amsterdam, pp. 5–14.

Tordoir, W. and Van Sittert, N.J. (1994) Organochlorines, in *Health Surveillance of Pesticide Workers: A Manual for Occupational Health Professionals*, W. Tordoir, M. Maroni, and F. He (Eds.), Elsevier, Amsterdam, pp. 51–57.

Van Sittert, N.J. and Tuinman, C.P. (1994) Coumarin derivatives (rodenticides), in *Health Surveillance of Pesticide Workers: A Manual for Occupational Health Professionals*, W. Tordoir, M. Maroni, and F. He (Eds.), Elsevier, Amsterdam, pp. 71–76.

WHO (1982) *Field Survey of Exposure to Pesticides: Standard Protocol*, VBC 82/1, World Health Organization, Geneva.

WHO (1984) *Environmental Health Criteria 39: Paraquat and Diquat*, World Health Organization/International Chemical Safety, Geneva.

WHO (1986) *Environmental Health Criteria 64: Carbamate Pesticides — A General Introduction*, World Health Organization/International Chemical Safety, Geneva.

WHO (1988) *Environmental Health Criteria 78: Dithiocarbamate Pesticides, Ethylenethio-urea, and Propylenethiourea — A General Introduction*, World Health Organization/International Chemical Safety, Geneva.

WHO (1989) *Environmental Health Criteria 82: Cypermethrin*, World Health Organization/International Chemical Safety, Geneva.

WHO (1994) *Environmental Health Criteria 153: Carbaryl*, World Health Organization/International Chemical Safety, Geneva.

Wray, J.E., Muftu, Y., and Dogramace, I. (1962) Hexachlorobenzene as a cause of porphyria turcica, *Turkish Journal of Pediatric*, 4:132–137.

Zielhuis, R.L. (1972) Epidemiological toxicology of pesticide exposure: report of an international workshop, *Archives of Environmental Health*, 25:399–405.

chapter two

Use of simultaneous biological monitoring and dermal dosimetry techniques to determine the exposure of chlorpyrifos to applicators and re-entry workers

R.C. Honeycutt, M. Honeycutt, M. DeGeare,
E.W. Day, Jr., B. Houtman, B. Chen, B.A. Shurdut,
R.J. Nolan, J.R. Vaccaro, P. Murphy, and M.J. Bartels

Contents

Introduction

The completion of a risk assessment for farmworkers who handle pesticides or who re-enter pesticide-treated cropland requires an array of data, including dermal deposition, dermal absorption, metabolism, pharmacokinetics, mammalian toxicology, and use pattern data. One of the most important elements of such a risk assessment is the estimation of the dermal deposition of the pesticide onto the skin of the farmworker.[1] Dermal exposure assessment of pesticides for farmworkers is commonly arrived at by two routes: (1) indirect measurement of skin exposure through biomonitoring[2] and (2) direct measurement of skin deposition using dosimetry techniques.[3]

The estimation of farmworker exposure using dermal dosimetry and urinary monitoring has been difficult to accomplish concurrently. This has been due to the assumption that patch or whole-body dosimeters appear to block the penetration of pesticide into the skin and in turn reduces the amount of pesticide or metabolite available for excretion. While seemingly not compatible to be performed concurrently, each method performed alone exhibits its own advantages; dermal dosimetry techniques allow for measurement of pesticides on the skin surface, while urinary monitoring permits measurement of the true body burden of an absorbed pesticide.

This paper summarizes a series of studies in which exposure to chlorpyrifos by mixer-loaders, applicators, and cleanup workers using air-blast or ground-boom spray equipment while handling Lorsban®* 4E Insecticide was determined utilizing concurrent whole-body dosimetry and biological monitoring techniques. These concurrent methods were also used to estimate chlorpyrifos exposure to workers re-entering both low and high crops after treatment with Lorsban® 4E Insecticide. The methods by which comparable results for dermal dosimetry and urinary monitoring were obtained are described.

---

* Registered trademark of DowElanco, Indianapolis, Indiana.

Materials and methods

Test materials and reference standards

The test material for this study was Lorsban® 4E Insecticide in 2.5-gallon plastic jugs, as well as Lorsban®* 50W in 5-pound bags. The reference standards were chlorpyrifos (O,O-diethyl-O-[3,5,6-trichloro-2-pyridinyl] phosphorothioate), CAS No. 002921-88-2; and 3,5,6-TCP (3,5,6-trichloro-2-pyridinyl).

The chlorpyrifos reference standard was used in conjunction with the analysis of coverall dosimeter sections (arms, legs, torso), air tubes, Gelman filters, head patches, and handwashes, while the 3,5,6-TCP was used as the reference standard during urine analysis.

Field controls and field fortifications

Field fortifications were prepared to check the field/storage stability of the dermal dosimeters, handwashes, and air filters. The field fortifications were prepared using the formulated product undiluted for "high" level spikes and diluted with water (~1 µg/mL chlorpyrifos) for the "low" level field spikes. Field fortification solutions for urine were prepared from a 3,5,6-TCP standard in acetonitrile utilizing an ~1.2-µg/mL solution for the "high" field fortifications and an ~0.01-µg/mL solution for the "low" level fortifications.

Field controls and field fortification samples were set up at each site to measure background levels of chlorpyrifos and to establish the stability of the analyte on sampling substrates under the environmental conditions at each site. For each application, two coveralls (subsequently cut into sections — arms, legs, torso — after exposure to the site environmental conditions), two sets of t-shirts and briefs, two pairs of head patches, and two air pumps with sampling trains were designated as field controls and placed on a clean table near the application site or re-entry site. At the same location, but on a separate table, duplicates of each substrate were set up and spiked with a "low" amount of chlorpyrifos. Duplicates of each substrate were also spiked with a "high" amount of chlorpyrifos. Air sampling tubes and Gelman filters were spiked with only a "low" amount of chlorpyrifos. All controls and fortified samples were exposed to the environment for 6 to 8 hours during the experimental run. Control and recovery handwashes were exposed for only 20 to 30 minutes.

Field control and field-fortified urine samples were prepared with urine from individuals with no known exposure to chlorpyrifos. Some control urine samples were left unfortified (controls), and some control urine samples were spiked with a "low" amount of 3,5,6-TCP, shaken, and stored on dry ice until analysis. Some control samples were also fortified with a "high" amount of 3,5,6-TCP.

---

* Registered trademark of DowElanco, Indianapolis, IN.

Monitoring equipment

Whole-body dosimetry, handwashes, head patches, and air sampling were performed on the day of application (or re-entry). Dermal dosimetry measurements were performed using outside whole-body dosimeters (coveralls), as well as inner-body dosimeters (t-shirts and briefs). Using this clothing scenario, blockage of chlorpyrifos penetration by the clothing is that which would normally occur under these work conditions. Abnormal blockage of penetration of chlorpyrifos to the skin would occur if, for example, patches were attached to the coverall to capture chlorpyrifos for the estimation of dermal exposure. This study design allows for concurrent dermal dosimetry and urinary monitoring measurements of worker exposure.

Head, neck, and hand exposures were measured using methods outlined in the literature.[4] Head patches were used to estimate dermal exposure to the neck and face of the worker. Handwashes were conducted using a 0.008% DSS solution and collected in 2-L Pyrex bowls. The handwash was repeated with distilled water, and the two handwash solutions were combined. The pooled handwash was then partitioned with ethyl acetate to remove the chlorpyrifos from the aqueous phase. An aliquot of the ethyl acetate was shipped to the analytical laboratory for analysis of chlorpyrifos.

Urine samples were collected in 2-L polypropylene containers on the day prior to application of chlorpyrifos, on the day of application (or re-entry), and for 4 days after application (or re-entry).

Air sampling was performed on the day of application (or re-entry) using personal air pumps fitted with a sampling train which consisted of a chromosorb air sampling tube and a Gelman filter connected to the pump in series with tygon tubing. The air samplers were calibrated to pull air through the sampling train at ~1 L/minute.

Analytical methods

Analysis of passive dosimeters

The analytical methods for the determination of chlorpyrifos in various substrates (passive dosimeters) have been described previously.[4-7] The two sections of the air tube were separated and each section placed in a vial with 5 mL hexane. The sample was shaken for 1 hr, and an aliquot of the extract was subjected to capillary gas chromatography using an electron capture detector (GC/ECD). The Gelman filter was extracted and analyzed the same way. The ethyl acetate extracts of the handwash solutions were analyzed by GC/ECD without any further processing. The denim patches, as well as the t-shirt and briefs samples, were extracted (shaken for 1 to 3 hr) with high-performance liquid chromatography (HPLC) grade iso-octane in the storage containers in which they came from the field. The extract was then analyzed by GC/ECD. The limits of detection ranged from 0.02 to 30 µg/mL for most analytical sets. Details of the analytical methods for passive dosimeters can be found in unpublished reports.[5-7]

Urine analysis for 3,5,6-TCP and creatinine

The urine samples were analyzed using a modified version of a published method.[8] The method involved fortification of the urine samples with an internal standard 3,4,5-trichloro-2-pyridinyl, which is a structural isomer of the 3,5,6-TCP metabolite of chlorpyrifos; hydrolysis of labile acid conjugates to 3,5,6-TCP; solvent extraction; derivitization to the *t*-butyl-dimethylsilyl ester of 3,5,6-TCP; and subsequent negative-ion chemical ionization gas chromatography/mass spectrometry (GC/MS) analysis. Creatinine was determined in urine using a modification of a method of Fabiny and Ertingshausen.[9]

Calculations of total respiratory and dermal exposure as well as calculation of chlorpyrifos body burden from urine data

Calculation of the dermal exposure, respiratory exposure, and total dermal absorbed dose for each volunteer was performed using a method by Honeycutt et al.[4] This was achieved by essentially calculating the amount of chlorpyrifos reaching the body surface through the clothing, correcting for dermal absorption of chlorpyrifos, and adding the amount of chlorpyrifos encountered through the respiratory route during the test. This value is called the total dermal absorbed dose (TDAD). The amount of exposure to chlorpyrifos based on urinary monitoring data (the body burden) was also calculated as shown in Nolan et al.[3] and Honeycutt et al.[4]

Calculation of total dermal absorbed dose

Total dermal exposure (TDE) was calculated using the following steps:

1. Determine the amount of chlorpyrifos on the outside torso section of the coverall and the t-shirt and briefs sample.
2. Determine the percent penetration of chlorpyrifos through the outer layer of the torso section of the coverall (micrograms chlorpyrifos on t-shirt and briefs)/(micrograms chlorpyrifos on torso section).
3. Sum the micrograms chlorpyrifos on the arms and legs and torso section of the coverall (outer whole-body dosimeter) and multiply by the percentage penetration of the chlorpyrifos through the coverall.
4. Add to this value the micrograms chlorpyrifos found in the handwash and head patches (corrected for surface area of the head and neck). This sum in terms of μg/kg body weight per day represents the total dermal exposure (TDE).
5. Correct for body weight by dividing the TDE by 70 kg (theoretical weight of adult). Correct for dermal absorption of 3% for chlorpyrifos penetration of the skin.

Total respiratory exposure (TRE) was calculated as follows:

1. Determine total chlorpyrifos (mg) on the air tube and Gelman filter.
2. Determine the volume of air passing through the air pumps during the test period. Divide the total µg chlorpyrifos on the air tube and Gelman filter by the total volume of air passing through the pumps. Correct for the active ventilation rate of 1.5 m^3/hr.
3. Correct for the work time during the test and correct for the 70-kg weight of an adult.
4. Assume 100% respiratory absorption of chlorpyrifos except for mixer-loaders, applicators, and cleanup workers in California who wore respirators which blocked 100% of the respiratory exposure.
5. This final value in terms of µg/kg body weight per day is the total respiratory exposure (TRE).

The total dermal absorbed dose (TDAD) = TRE + TDE.

Calculation of the chlorpyrifos body burden

The absorbed dose or body burden of chlorpyrifos can be estimated using urinary monitoring data as follows:[3]

1. Normalize the 3,5,6-TCP in the urine specimen: mg 3,5,6-TCP in urine × 1.8 g creatinine per day divided by observed creatinine excretion rate.
2. Fit the amount of 3,5,6-TCP excreted in individual urine samples to a one-compartment pharmacokinetic model that describes the time course of 3,5,6-TCP in urine of volunteers following the application of chlorpyrifos to the forearm (72% of dose excreted).[3]

Study design-description of the field phase of the study

This series of studies describes the use of both dermal dosimetry and urinary monitoring techniques to determine the exposure to applicators, mixer-loaders, cleanup workers, re-entry scouts, re-entry harvesters, and re-entry tree pruners who were exposed to chlorpyrifos through the performance of typical agricultural tasks. Table 1 describes the various farmworker exposure scenarios examined during this study. Both liquid Lorsban® 4E and wettable powder Lorsban® 50W were used as test substances. The studies were performed in four states (California, Michigan, Florida, and Arizona) using oranges and lemons (California), bare ground and Christmas trees (Michigan), cauliflower (Arizona), and tomatoes (Florida) as crops while using ground-boom sprayers (Michigan, Arizona, and Florida) and air-blast sprayers (California).

Table 2 represents a summary of the design of the chlorpyrifos re-entry studies. Five sites (three in California, one in Arizona, and one in Florida) were selected. The California orange harvesting and lemon tree pruning represented high crop re-entry activities, while the Florida tomato and

Table 1 Field Phase of Chlorpyrifos Worker Exposure and Re-entry Studies

	Descriptions			
	Applicator/mixer-loader		Re-entry	
Chemical	Lorsban® 4E	Lorsban® 50W	Lorsban® 4E	Lorsban® 50W
Crop location	CA, MI	FL, AZ	CA	FL, AZ
Crops	Oranges Lemons Bare ground Christmas trees	Tomatoes Cauliflower	Oranges Lemons	Tomatoes Cauliflower
Rates of application (lb/a.i./A)	1–7	1	6	1
Types of workers	Mixer-loaders Applicators Cleanup workers	Mixer-loaders Applicators	Citrus pickers Citrus pruners	Scouts
Acres worked	10–61	23–43	1–2	—
Re-entry interval (days)	—	—	2, 43	1

Table 2 Chlorpyrifos Re-entry Study Parameters

Location	5 sites (3 CA, 1 AZ, 1 FL)
Volunteers	CA site no. 1 = 5 orange pickers CA site no. 2 = 5 orange tree pruners CA site no. 3 = 5 lemon tree pruners FL site = 5 tomato scouts AZ site = 5 cauliflower scouts
Re-entry intervals	CA site no. 1 = 43 days CA site nos. 1 and 2 = 2 days FL and AZ sites = 1 day
Re-entry work period	CA = 5–6 hr; FL and AZ = 4 hr
Dosimetry	CA, outer whole-body dosimeter = coveralls dosimeters (arms, legs, torso); FL and AZ (short sleeve with arm band); inner dosimeter = t-shirt and brief; head patch, face and neck
Handwash	Before re-entry Lunch break After re-entry
Respiratory	Chromosorb air sorbent tubes (vapors) Gelman filters (particulates and aerosol)
Biomonitoring	Urine: Collect ~12-hr samples over a 6-day period (–1, 0, 1, 2, 3, 4 days)

Table 3 Chlorpyrifos Mixer-Loader, Applicator Exposure Study Parameters

Location	CA, MI, FL, AZ
Rates of application (lb/a.i./acre)	CA = 5–7, MI = 1.6–2.8, FL = 2.0, AZ = 0.9–1.5
Volume of spray/acre	CA = 500–700 gal/acre, MI = 10–30 gal/acre, FL = 60–120 gal/acre, AZ = 38–46 gal/acre
Volunteers	CA: 15 mixer-loaders (citrus); 15 applicators (air-blast); 15 cleanup workers. MI: 3 mixer-loaders and 3 applicators. FL: 3 mixer-loaders and 3 applicators. AZ: 3 mixer-loaders and 3 applicators.
Volunteer work period	CA = 6 hr; MI, FL, AZ = 4–5 hr
Dosimetry	CA, outer coveralls (arms, legs, torso) + short-sleeved shirt + pants + t-shirt and briefs, head patch, and handwash. MI, FL, AZ, outer coveralls (arms, legs, torso) + t-shirt and briefs, head patch, and handwash
Respiratory	Chromosorb air tubes and Gelman filters, air pump
Biomonitoring	Urine collection of ~12-hr samples over a 6-day period (–1, 0, 1, 2, 3, 4 days)

Arizona cauliflower scouting represented low crop activities. Five orange harvesters and five lemon tree pruners were tested using concurrent dosimetry and urinary monitoring techniques. The re-entry intervals were 43, 2, and 1 days for the orange harvesters, lemon tree pruners, and scouts, respectively. Re-entry workers were monitored for 5- to 6-hr work days in California and for a 4-hr work day in Florida and Arizona.

Table 3 presents a summary of the design of the mixer-loader, applicator, and cleanup worker portion of this study. Mixer-loaders, applicators, and spray equipment cleanup workers were monitored using concurrent dosimetry and urine collection techniques in four states. Fifteen mixer-loaders, 15 air-blast applicators, and 15 cleanup workers were monitored in California. Three mixer-loaders and three applicators were tested in each of three states (Florida, Arizona, and Michigan) while handling Lorsban® 4E and Lorsban® 50W during ground-boom application to Christmas trees, bare ground, tomatoes, and cauliflower. The application rates ranged from 1 to 7 lb active ingredient (a.i.) per acre and test periods ranged from 4 to 5 hr in Michigan, Florida, and Arizona to 6 hr in California.

Results

Field controls

Small amounts of chlorpyrifos were found in a number of control samples. The levels observed were probably dependent on the location of the control samples relative to the spray site. The observed background levels of

Table 4 Field Controls and Field Spikes (Worker Exposure)

Matrix	"Low" spike (μg chlorpyrifos)		"High" spike (μg chlorpyrifos)	
	CA[a]	MI, FL, AZ	CA[a]	MI, FL, AZ
Coverall sections	1198	40.2	191,720	166,400
T-shirt and briefs	1198	10.2	47,930	166,400
Handwash[b]	120	12.1	47,930	12,400
Head patches	119–1198	5.1	47,930	1456
Air tubes	11–36	0.1–0.5	—	1–5
Gelman filters	11–36	0.1–0.5	—	1–5

[a] 1992.

[b] Handwash = two washes in handbowl, 250 mL each (0.008% DSS).

chlorpyrifos in field controls were subtracted from the amounts of chlorpyrifos found in the corresponding field fortification samples for the purpose of calculating field recoveries.

Low levels of 3,5,6-TCP were also observed in pre-exposure urine from most of the field workers. These 3,5,6-TCP levels were subtracted from urine field recovery samples and were used to correct levels of 3,5,6-TCP found in the post-exposure urine samples from these same workers. This procedure was necessary to calculate the amount of 3,5,6-TCP in the urine that was attributable to the exposure period.

Field recoveries (field fortifications)

Table 4 presents the amounts of chlorpyrifos used to fortify the various substrates in this study. Chlorpyrifos "low" spike levels for coverall sections (arms, legs, and torso) as well as t-shirts and briefs, handwashes, and head patches ranged from 5.1 μg to 1198 μg. "High" spike levels of these substrates ranged form 1456 μg to 191,720 μg. Air tubes and Gelman filters were fortified with only "low" amounts of chlorpyrifos which ranged from 0.1 μg to 36 μg.

Table 5 presents data on field spike recoveries. Recoveries of chlorpyrifos from "low" spiked substrates in the California studies ranged from 62.5 to 126%, and recoveries of chlorpyrifos from "high" spiked substrates ranged from 93.7 to 133%. In the Florida, Arizona, and Michigan studies, all recoveries ("low" and "high") ranged from 61 to 158%. The field recoveries cited above were found to be reasonable and within the range of field recoveries seen in many worker exposure studies.

Deposition of chlorpyrifos on passive dosimeters

The amount of chlorpyrifos deposited on the passive dosimeters such as outer whole-body dosimeters (coveralls) was determined as described

Table 5 Chlorpyrifos Field Spike Recoveries,
Chlorpyrifos Worker Exposure Study

| | Percent Recovery | | |
| | CA (1992) | | MI, FL, AZ |
Matrix	Low	High	Low + High
Air tubes	82.6 ± 19.4	—	75–106
	(n = 24)		
Gelman filters	80.4 ± 15.6	—	98–108
	(n = 24)		
Coveralls sections (A)	82.7 ± 23.5	107 ± 10.5	61–103
	(n = 18)	(n = 24)	
T-shirt and briefs	69.9 ± 10.6	98.8 ± 13.4	83–100
	(n = 18)	(n = 24)	
Handwashes	126 ± 16.8	133 ± 22.7	85–105
	(n = 17)	(n = 3)	
Head patches	62.5 ± 13.8	103 ± 19.7	79–93
	(n = 16)	(n = 22)	
Urine	120 ± 36	93.7 ± 50	104–158
	(n = 35)	(n = 37)	

above. The coveralls were sectioned into arm, leg, and torso samples (two arms equaled one sample; two legs equaled one sample). Amounts of chlorpyrifos deposited on each coverall section were recorded. Details of the amount of chlorpyrifos deposited on the passive dosimeters including coverall sections, head patches, air tubes, Gelman filters, t-shirt and briefs, and handwashes can be found in unpublished reports by Honeycutt et al. and Shurdut et al.[5-7]

Penetration of chlorpyrifos through the outer whole-body dosimeter to the inner-body dosimeter

In order to determine the dermal exposure of volunteers to chlorpyrifos, the penetration of chlorpyrifos through the outer whole-body dosimeter (coveralls) to the inner body dosimeter (t-shirt and briefs) was measured. The penetration factor was calculated for each volunteer in the study from the experimental data by dividing the amount of chlorpyrifos on the t-shirt and brief sample by the amount of chlorpyrifos on the torso section of the coveralls. This method of calculation assumes that the surface area of the torso section of the coveralls is nearly the same as the surface area of the t-shirt and briefs worn directly under the torso section of the coveralls. A mean penetration factor for each worker type was calculated by averaging all the worker volunteer

Table 6 Penentration Factors Calculated for Chlorpyrifos
Worker Exposure and Re-entry Studies

Worker type	No. replicates	Location	Mean penetration factor
Applicators	9	MI, FL, AZ	7.0
Mixer-loaders	9	MI, FL, AZ	7.5
Scouts	1	FL, AZ	8.9
Applicators	9	CA	4.8
Mixer-loaders	9	CA	5.3
Cleanup workers	9	CA	12.4
Citrus pruners	5	CA	5.0
Citrus pruners	5	CA	12.3

Note: One layer of clothing was penetrated.

penetration factors within one worker group (e.g., California mixer-loaders). Table 6 shows the mean penetration factors for all worker types in this study. The mean penetration factor ranged from 4.8 to 12.4% (overall mean = 7.9%) for penetration through one layer of clothing. These values reflect penetration factors for one layer of clothing found in previous studies.[10,11]

Comparison of exposure levels — total absorbed dose of chlorpyrifos (passive dosimetry vs. urinary monitoring)

The total absorbed dose of chlorpyrifos for each volunteer in this series of worker exposure and re-entry studies was calculated using either passive dosimetry or urinary monitoring data. Table 7 shows a comparison of estimated chlorpyrifos levels for each type of worker volunteer calculated either from urinary data or passive dosimetry data. The geometric means as well as the arithmetic means of exposure in terms of mg/kg body weight per day for each worker type are shown in Table 7. Comparison of the arithmetic mean exposure levels using passive dosimetry or urine monitoring methods shows excellent agreement. Comparison of the geometric means of the exposure data also shows good correlation between the two exposure assessment methods.

Figures 1 and 2 show the correlation of exposure results using dosimetry and urine monitoring data. Plotting the arithmetic means of exposure data using biomonitoring against the arithmetic means of exposure data using dosimetry shows good correlation, with $r = 0.93$. Plotting geometric means of the same data also shows a good correlation between the passive dosimetry results and the biomonitoring results, with $r = 0.89$.

Figure 3 shows graphically the comparison of the means of chlorpyrifos exposure levels arrived at using passive dosimetry vs. biomonitoring exposure assessment methods. Again, it is seen that the two methods compare well in terms of total exposure for any given worker type. Exposure appears to be a function of worker task, with equipment cleanup tasks and re-entry work being at the low end of the chlorpyrifos exposure spectrum and mixer-loaders at the high end.

Table 7 Comparison of Exposure Estimated Levels (Total Absorbed Dose of Chlorpyrifos) Dosimetry vs. Urine Monitoring

| Worker type | Crop/ location | No. replicates | Total absorbed dose of chlorpyrifos (μg/kg bw/day) | | | |
| | | | Dosimetry | | Urine monitoring | |
			Arithmetic mean	Geometric mean	Arithmetic mean	Geometric mean
Mixer-loader	Citrus/CA	15	3.0	1.8	5.0	4.2
Applicator	Citrus/CA	15	5.9	3.5	5.7	3.6
Cleanup worker	Citrus/CA	15	0.51	0.34	1.0	0.85
Applicator (4E)	Small trees, bare ground/ MI	3	1.4	1.5	1.9	1.3
Mixer-loader (4E)	Small trees, bare ground/ MI	3	9.6	6.3	7.8	7.5
Applicator (50W)	Tomatoes, cauliflower/ FL, AZ	6	2.3	1.5	2.3	1.3
Mixer-loader (50W)	Tomatoes, cauliflower/ FL, AZ	6	12.8	6.3	11.7	7.5
Citrus pickers	Citrus/CA	5	0.25	0.23	0.78	0.73
Citrus pruners	Citrus/CA	5[a]	2.5	2.5	6.4	6.3
		5[a]	1.0	1.0	3.0	2.9
Scouts	Tomatoes, cauliflower/ FL, AZ	10	0.9	0.7	1.5	1.1

[a] Wet.

Conclusions

1. Laboratory analytical recoveries and field spike recoveries were acceptable for all substrates encountered in this series of studies. Calculated penetration factors were similar to penetration factors reported in the literature.
2. The described method of simultaneously measuring the total absorbed dose by passive dosimetry and biomonitoring shows excellent reproducibility over a range of four states, six worker types, and a range of 84 test replicates (one replicate = one volunteer).

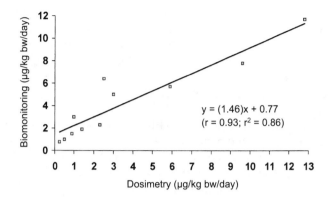

Figure 1   Correlation of dosimetry and urine monitoring using arithmetic means.

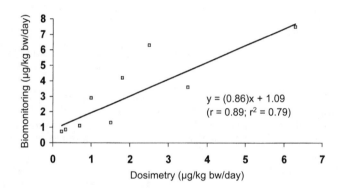

Figure 2   Correlation of dosimetry and urine monitoring using geometric means.

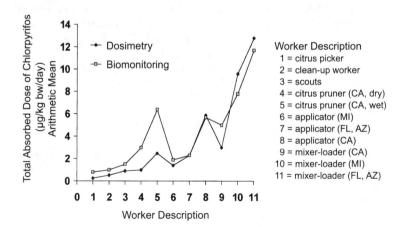

Figure 3   Total absorbed dose of chlorphyrifos.

3. The methods described are precise enough to show quantifiable differences in exposure for different worker types and worker tasks.

4. The data gathered during these series of experiments indicate that concurrent passive dosimetry and biomonitoring methods are both precise and accurate enough to provide a true range of exposure values for various types of workers handling agrochemicals or re-entering treated crops.

5. Collectively, the data from Table 7 and Figures 1 through 3 lead to the conclusion that concurrent biomonitoring and passive dosimetry techniques can be achieved and are not divergent worker exposure assessment methods. The correlation between exposure levels measured by these methods is quite good.

References

1. Wolfe, H.R., Durham, W.F., and Armstrong, J.F., *Archives Environmental Health*, 14, 622, 1967.

2. Durham, W.F. and Wolfe, H.R., *Bulletin WHO*, 26, 75–91, 1962.

3. Nolan, R.J., Rick, D.L., Freshour, N.L., and Saunders, J.H., Chlorpyrifos: pharmacokinetics in human volunteers, *Toxicology and Applied Pharmacology*, 73, 8–15, 1984.

4. Honeycutt, R.C., Honeycutt, M.C., Vaccaro, J.R., Murphy, P.G., Bartels, M.J., Nolan, R.J., and Day, E.W., unpublished data, 1997.

5. Honeycutt, R.C. and DeGeare, M.A., Worker Reentry Exposure to Chlorpyrifos in Citrus Treated with Lorsban® 4E Insecticide, unpublished report, HERAC, Inc., December 10,1993.

6. Honeycutt, R.C. and DeGeare, M.A., Evaluation of the Potential Exposure to Chlorpyrifos During Mixing and Loading, Spray Application, Clean-Up Procedures During the Treatment of Citrus Groves with Lorsban® 4E Insecticide, unpublished report, HERAC, Inc., January 18, 1994.

7. Shurdut, B.A., Murphy, P.G., Nolan, R.J., and McNett, D.A., Lorsban® 4E and 50W Insecticides: Assessment of Chlorpyrifos Exposures to Applicators, Mixer/Loaders, and Re-entry Personnel During and Following Application to Low Crops, DowElanco Laboratory Project DECO-HEH 2.1-1-182(118) and DECO-HEH2.1-1-182(124), September 27, 1993.

8. Bartels, M.J. and Kastl, P.E., Analysis of 3,5,6-trichlopyridinol in human urine using negative-ion chemical ionization gas chromatography-mass spectrometry, *Journal of Chromatography*, 575, 69–74, 1992.

9. Fabiny, P.L. and Ertingshausen, E., Automated reaction rates method for determination of serum creatinine with the centrifichem, *Clinical Chemistry*, 17, 696–700, 1971.

10. Kurtz, D.A. and Bode, W.M., Application exposure to the home gardener, in *Dermal Exposure Related to Pesticide Use: Discussion of Risk Assessment*, ACS Symposium Series 273, Honeycutt, R.C., Zweig, G., and Ragsdale, N.N., Eds., American Chemical Society, Washington, D.C., 1985, pp. 139–161.

11. *Pesticide Assessment Guidelines*, Subdivision U, *Applicator Exposure Monitoring*, U.S. Environmental Protection Agency, Washington, D.C., 1986.

chapter three

Use of probability and distributional analysis of chlorpyrifos worker exposure data for the assessment of risks

E.W. Day, Jr., W.L. Chen, K.D. Schnelle, and B.A. Shurdut

Contents

Summary

Assessments of risks associated with the use of chlorpyrifos insecticide products for workers have been made. The assessments are based on the results of field studies conducted in citrus groves, a Christmas tree farm, cauliflower and tomato fields, and greenhouses that utilized both passive dosimetry and biomonitoring techniques to determine exposure. The biomonitoring results likely provide the best estimate of absorbed dose of chlorpyrifos, and these have been compared to the acute and chronic no observed effect levels (NOELs) for chlorpyrifos. Standard margin-of-exposure (MOE) calculations using the geometric mean of the data are performed; however, probability (Student's *t*-test) and distributional (Monte Carlo simulation) analyses are deemed to provide more realistic evaluations of exposure and risk to the exposed population.

Introduction

Risk assessment pertains to characterization of the probability of adverse health effects occurring as a result of human exposure. Recent trends in risk assessment have encouraged the use of realistic exposure scenarios, the totality of available data, and the uncertainty in the data, as well as their quality, in arriving at a best estimate of the risk to exposed populations. The use of "worst case" and even other single point values is an extremely conservative approach and does not offer realistic characterization of risk. Even the use of arithmetic mean values obtained under maximum use conditions may be considered to be conservative and not descriptive of the range of exposures experienced by workers. Use of the entirety of data is more scientific and statistically defensible and would provide a distribution of plausible values.

The previous chapter presented results from field studies in which exposures were measured for workers involved with the use of the insecticide chlorpyrifos in several use scenarios and for persons who might re-enter treated areas. In this chapter, the results from these studies are handled by several methods to demonstrate the advantages of using probability and distributional analyses, rather than single point values, for the characterization of risks to pesticides.

Toxicity endpoint

The purpose of this chapter is not to discuss the merits, or lack thereof, of using plasma cholinesterase inhibition as an adverse effect in quantitative risk assessments for chlorpyrifos or other organophosphate pesticides. A number of regulatory agencies consider the inhibition of plasma cholinesterase to be an indicator of exposure, not of toxicity. The U.S. Environmental Protection Agency, at this point, continues to use this effect as the basis for calculating the reference doses for chlorpyrifos, and it is thus used here for assessing risks.

Based on the data from controlled human studies, the NOEL for plasma cholinesterase inhibition for a single dose of chlorpyrifos is between 0.1 and 0.5 mg/kg bw/day, and the more conservative 0.1 mg/kg bw/day (100 µg/kg bw/day) is used in this assessment as the acute NOEL for chlorpyrifos. The repeated dose NOEL in humans is 0.03 mg/kg bw/day (30 µg/kg bw/day), based on plasma cholinesterase activity, and this is the basis for the establishment of the reference dose of 0.003 mg/kg bw/day (3 µg/kg bw/day) used by the EPA in assessing dietary risk to chlorpyrifos. For the work described here, both NOELs are used as bases for assessing risks to persons who have the potential for non-dietary exposure to chlorpyrifos. For exposures that are infrequent or of short duration, the 100 µg/kg bw/day NOEL is assumed to be the more appropriate value, and the lower 30 µg/kg bw/day will be used in those situations in which exposure may be considered to be more frequent.*

Types of analyses

Three approaches to risk analysis will be presented here for the available chlorpyrifos exposure data, namely: (1) the single point, margin of safety approach; (2) probability analysis; and (3) Monte Carlo simulation.

Single point

The most commonly used approach, and most conservative, for assessing risks to humans is to compare the exposure results directly to the appropriate toxicological endpoint by calculating the margin of exposure (MOE; margin of safety in many countries), which is given by:

$$MOE = \frac{NOEL}{Exposure}$$

An exposure value may be calculated for each worker in a study, or the arithmetic or geometric mean of all of the participants may be used to estimate the central tendency of the entire data set. For studies in which there is a wide range of results, the geometric mean is deemed to be the more appropriate for estimating the central tendency of exposure values over a range of exposed persons. The geometric mean is generally used in this presentation, although the arithmetic mean results will be presented for purposes of comparison.

Probability

The second method involves determination of the probability of an exposure exceeding a known toxicological endpoint. If it is assumed that the absorbed

---

* In September 1999, the United Nations' World Health Organization (WHO) and the Food and Agriculture Organization (FAO) Joint Meeting on Pesticide Residues (JMPR) established an acute reference dose (R_fD) of 0.1 mg/kg bw and a chronic R_fD, or acceptable daily intake (ADI), of 0.01 mg/kg bw/day. (See FAO/WHO Report 153, *Pesticide Residues in Food*, Section 4.7, Chlorpyrifos, 1999.)

doses measured for the individuals monitored in a study can be characterized within a larger distributional framework, the probability of exceeding
a particular dose level can be adequately described. It is generally recognized
that, given an infinite number of exposure points, exposure data would
closely follow a lognormal distribution. However, due to the inherent limitations of the data, both the lognormal and *t*-distributions are used in this
analysis. The *t*-distribution is a better statistical analysis than fitting a Gaus-
sian or lognormal distribution when the number of samples is small (n < 30)
and when the population from which the sample is taken is assumed to be
normally distributed. Consequently, a *t*-distribution was fit to both the actual
estimated doses and the lognormally transformed doses in order to determine the probability of exceeding the selected NOEL.

Monte Carlo simulation

The third method used to interpret the level of risk associated with chlorpyrifos use is Monte Carlo simulation. This method provides a range of exposure
estimates for the evaluation of the uncertainty in a risk estimate based on
ranges of input variables. The first step in performing a Monte Carlo simulation is determination of a model to describe the dose. This model describes
the relationship between the input parameters and dose, and a specific model
is presented here for one group of workers.

The second step in conducting a Monte Carlo simulation is to identify
the distribution inputs for those factors that are allowed to vary. Frequency
distributions are utilized in this model for most of the input parameters,
while some factors are held constant if deemed appropriate.

The third step is to select the number of iterations or calculations of dose
that are to be performed as a part of each simulation. For the analysis here,
a total of 10,000 iterations based on the selection of input variables from each
defined distribution were performed as part of each simulation. The large
number of iterations performed, as well as the Latin hypercube sampling
(non-random sampling) technique employed by the Crystal Ball® simulation
program, ensured that the input distributions were well characterized, that
all portions of the distribution (such as the tails) were included in the analysis, and that the resulting exposure distributions were stable.

Results and discussion

Study results

The results from the several studies that have been conducted to measure
exposures associated with the use of chlorpyrifos are summarized in Tables 1
and 2. Table 1 summarizes results from mixer-loader and applicator studies
reported by Honeycutt et al.[1] Listed for each work description are the number
of replicates, the arithmetic mean, and the geometric mean for the replicates
from both the passive dosimetry measurements and the biomonitoring tech-

Table 1 Summary of Estimated Absorbed Doses from Mixer-Loader
and Applicator Studies with Chlorpyrifos

		Passive dosimetry (µg/kg/day)			Biological monitoring (µg/kg/day)	
Description	n	Arithmetic mean	Geometric mean	n	Arithmetic mean	Geometric mean
M/L of 4 lb/ gal EC	18	6.3	2.9	18	10.4	4.3
Applicators, air-blast to citrus	15	6.8	4.0	13	5.7	3.7
M/L, 50 WP	3	13.3	12.7	6	16.5	13.8
Applicators, ground-boom to low crops	9	2.7	2.1	9	2.8	2.2
M/L/A in greenhouses, microencap-sulated	16	3.1	0.7	16	1.0	0.5
Clean-up, air-blast equipment, 4 EC	15	0.6	0.4	15	1.0	0.8

Note: M/L = mixing/loading; EC = emulsifiable concentrate; 50 WP = 50% wettable power; A = application.

nique. For some of the work patterns, there is little difference between the arithmetic and geometric means, which is indicative of a relatively narrow range of values and few, if any, outliers. Where there is significant discrepancy between the two means, examination of the individual data points shows the presence of 1 or 2 outliers that skew the arithmetic mean to higher values. It is particularly important to use the geometric mean in the latter case to prevent one especially high or low value from having too great an influence on the central tendency value to be used in the margin of exposure (MOE) calculation.

Table 2 summarizes the re-entry exposure data from studies with chlorpyrifos.[1] There are fewer replicates for these workers which would seem to be justified by the lower variability in the data sets. There are practically no differences between the arithmetic and geometric means for these data sets.

Margin of exposure (safety)

The MOE calculations for the various exposure scenarios with chlorpyrifos are presented in Table 3. Listed in this table are the calculated results for the workers involved with mixing and loading and the application of chlorpyrifos,

Table 2 Summary of Estimated Absorbed Doses for Re-entry
Activities in Areas Treated with Chlorpyrifos

Description	n	Passive dosimetry (μg/kg/day)		n	Biological monitoring (μg/kg/day)	
		Arithmetic mean	Geometric mean		Arithmetic mean	Geometric mean
Citrus picker, 43-DAT	5	0.25	0.23	5	0.78	0.73
Citrus pruner, wet foliage, 2-DAT	5	2.5	2.5	5	6.4	6.3
Citrus pruner, dry foliage, 2-DAT	5	1.1	1.1	5	3.0	2.9
Scouts, low crop, 1-DAT	10	1.3	.12	10	2.9	2.1

Note: DAT = days after treatment.

as well as for workers who re-enter groves or fields treated with chlorpyrifos. Included in this table for each scenario is the geometric mean of the exposure results from the biomonitoring phase of the studies and the calculated margins of exposure using both the 30 and the 100 μg/kg bw/day NOELs. Note that the acute NOEL of 100 μg/kg bw/day is also the chronic LOEL (lowest observed effect level); there are some who feel that the LOEL should be used as the benchmark for plasma cholinesterase activity instead of the NOEL, as this effect should be considered an indication of exposure, not of toxicity.

Note that one of the MOEs for each scenario is underlined. This is the value that is deemed to be the more appropriate based on the frequency and duration of activity. For example, mixer-loaders and applicators in orchards and low crops may be considered to experience chronic exposure as these workers, particularly those who are custom applicators, may handle chlorpyrifos products repeatedly during periods of several weeks in length. The same may be said for harvesters, pruners, and scouts who may have repeated exposure during certain periods of the year. In contrast, after a greenhouse worker applies chlorpyrifos, the individual is not likely to handle the product again until the next application, which could be several weeks. Because chlorpyrifos is relatively rapidly metabolized and dissipated in the human body (half-life of less than 2 days), the single-dose NOEL is the more appropriate toxicity endpoint for calculating the MOE. For those who consider the LOEL as the more relevant dose value, the results in the "100 μg/kg bw/day" column would be the more appropriate MOE values to use for regulatory decisions.

The calculated MOEs in Table 3 demonstrate some of the problems with using this approach for characterizing risks and making regulatory decisions. Several of the exposure scenarios have MOEs of less than 10, which is the

Table 3 Estimated Margins of Exposure for Chlorpyrifos from Biomonitoring Data Using Geometric Means (Single-Point Approach)[a]

Activity	Exposure (geometric mean) (μg/kg bw/day)	MOE vs. chronic NOEL (30 μg/kg bw/day)	MOE vs. acute NOEL (100 μg/kg bw/day)
Mixing and loading, 4 lb/gal EC	4.3	<u>7.0</u>	23
Mixing and loading, 50 WP	13.8	<u>2.2</u>	7.2
Air-blast application to citrus groves	3.7	<u>8.1</u>	27
Ground-boom application, low crop	2.2	<u>13.6</u>	46
Mix, load, and apply micro-encapsulated product in greenhouse	0.5	60	<u>200</u>
Cleanup of air-blast equipment, EC formulation	0.8	<u>38</u>	125
Citrus harvester, 43-DAT	0.7	<u>43</u>	143
Citrus pruner, wet foliage, 2-DAT	6.3	<u>4.8</u>	16
Citrus pruner, dry foliage, 2-DAT	2.9	<u>10.3</u>	35
Scouting, low crop, 1-DAT	2.1	<u>14.3</u>	48

[a] Underlined values are the more appropriate, based on the frequency and duration of activity.

Note: EC = emulsifiable concentrate; WP = wettable powder; DAT = days after treatment.

uncertainty factor deemed acceptable for compounds such as chlorpyrifos. Thus, regulators would have to require registrants to implement mitigation measures to reduce exposure and increase the MOE. Yet, in all of these studies, none of the workers was observed to have had plasma cholinesterase activity affected by the exposure, nor did any workers experience any clinical signs of toxicity. Of course, it would not be expected that workers would have toxic symptoms if the NOEL in humans is truly higher than the measured exposure doses. Nevertheless, the regulatory community should have more advanced methods of interpreting exposure results, such as statistical analysis, in order to characterize risks in a more realistic manner.

Statistical analysis

The results from using the Student's *t*-test for a distributional analysis are presented in Table 4. These results indicate the probability of a given worker in the listed scenario exceeding the NOEL of the toxicity endpoint. The probability of exceeding the LOEL and of thus experiencing a depression of plasma cholinesterase activity is not given (except for chronic exposure scenarios in the "100 µg/kg bw/day" column). Hence, even these probabilities may be considered to be conservative and not fully representative of the probability of a worker actually experiencing a toxic effect.

When the data in Table 4 were given full evaluation, it was recommended that the 50W formulation of chlorpyrifos no longer be marketed in bags that allowed significant exposure for mixer-loaders of this product. This product was removed from the marketplace and was replaced with one in which the wettable powder (WP) is in water-dissolvable packets. Exposure data on other active ingredients have clearly demonstrated reduced exposure with this type of packaging. The other uses were deemed to present a minimal hazard to users, and only minor protective measures have been recommended to workers.

The probabilities for re-entry situations presented in Table 4 indicate there is no realistic probability of an individual experiencing toxic symptoms or plasma cholinesterase activity depression when performing these activities following application of chlorpyrifos. In general, estimating the probability of individuals experiencing a toxic effect seems to be a more relevant method of characterizing risks to a given population than the use of the standard MOE calculation, which uses a single central tendency value.

Monte Carlo simulation

The third approach for estimating risk from field data is the use of Monte Carlo simulation. This has been done in this study for the greenhouse mixer/loader/applicators,[5] where the model that describes the relationship between the input variables is expressed by the following equation:

$$D = \frac{(DR \times T \times 0.03) + (C \times BR \times T)}{BW}$$

where:

D	=	average daily dose (mg/kg)
DR	=	dermal deposition rate (mg/hr)
T	=	duration of work period (hr)
C	=	respiratory concentration (mg/m³)
BR	=	breathing rate (m³/hr)
BW	=	weight of individual monitored (kg)

Table 4 Statistical (Probability) Analysis of Chlorpyrifos
Biomonitoring Data for Mixer-Loaders and Applicators

Activity	Probability of exposure exceeding chronic NOEL of 30 µg/kg bw/day	Probability of exposure exceeding acute NOEL of 100 µg/kg bw/day
Mixing and loading of 4 lb/gal EC	0.030 3 in 100	0.0029 3 in 1000
Mixing and loading of 50 WP	0.242 1 in 4	0.024 1 in 40
Air-blast application to citrus grove	0.021 2 in 100	0.0019 2 in 1000
Ground-boom application to low crops	0.0106 1 in 100	0.0012 1 in 1000
Mix, load, and apply microencapsulated product in a greenhouse	N/A[a]	0.0001 1 in 10,000
Cleanup of air-blast equipment	<0.000001 <1 in 1,000,000	<0.000001 <1 in 1,000,000
Citrus harvesting, 45-DAT	0.000021 2 in 100,000	0.000003 3 in 1,000,000
Citrus pruning, 2-DAT	0.0021 2 in 1000	0.00014 1 in 10,000
Scouting, low crops, 1-DAT	0.0060 6 in 1000	0.00066 7 in 10,000

[a] N/A = not analyzed, as exposure is short term.

Note: EC = emulsifiable concentrate; WP = wettable powder; DAT = days after treatment.

In this case, the frequency distributions were allowed to vary for all of the input parameters, except one. The factor of 0.03 remained fixed, as this represents the fraction of chlorpyrifos that is expected to be absorbed through the skin upon contact based upon dermal absorption studies.[2] The body weight distribution is shown in Figure 1, where the body-weight values were derived from published data sources for the working population and follow a normal distribution. The body weight was not allowed to go below 55 kg or above 105 kg. However, for the other input parameters (DR, C, and T), frequency distributions were generated given the ranges and statistical descriptors of the data collected in the study and the assumption of the shape of the distribution (normal vs. lognormal). For example, the dermal deposition rate (DR) distribution is shown in Figure 2. The parameters for the distribution were taken from the mean and standard deviation of the

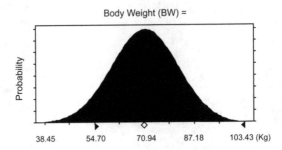

Normal distribution with parameters:
Mean   70.94
Standard Dev.   10.83

Selected range is from 55.00 to 105.00
Mean value in simulation was 72.48

Body Weight (BW) =

38.45   54.70   70.94   87.18   103.43 (Kg)

Figure 1 Monte Carlo simulation: body weight distribution.

observed data, and it was assumed that dermal exposures followed lognormal distributions. Because the biomonitoring data are limited in their ability to discern the individual components of dose, only the passive dosimetry results were used for the determination of frequency distributions.

The breathing rate data used to define the BR variable were adapted from the reported distribution generated from Shamoo et al.[3] In the Shamoo study, a different distribution was identified for several activity patterns, and for this simulation the slow, medium, and fast rate classifications were combined. The distribution is shown in Figure 3.

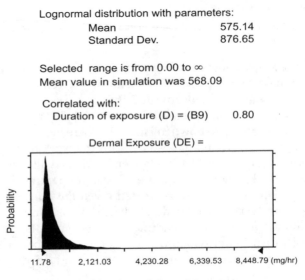

Lognormal distribution with parameters:
Mean   575.14
Standard Dev.   876.65

Selected range is from 0.00 to ∞
Mean value in simulation was 568.09

Correlated with:
Duration of exposure (D) = (B9)   0.80

Dermal Exposure (DE) =

11.78   2,121.03   4,230.28   6,339.53   8,448.79 (mg/hr)

Figure 2 Monte Carlo simulation: dermal deposition rate.

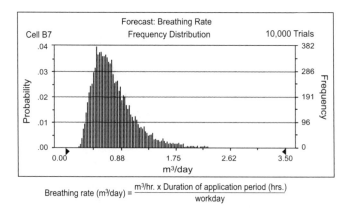

$$\text{Breathing rate (m}^3/\text{day)} = \frac{\text{m}^3/\text{hr. x Duration of application period (hrs.)}}{\text{workday}}$$

Figure 3 Monte Carlo simulation: breathing rate distribution.

Means and standard deviations for these distributions were normalized to daily breathing rates (m³/day), and an acceptable range was defined. It was assumed that the "day" represents the duration of time within a working day that chlorpyrifos may be handled by an individual (0.25 to 6.0 hr). It was also assumed that exposures would be negligible for the remainder of the working day following application or other contact. Both the dermal and inhalation exposures were assumed to follow lognormal distributions, which is consistent with common practice for exposure data distributions (for example, in the Pesticide Handlers Exposure Database, PHED).

A total of 10,000 iterations or calculations of dose were performed as part of this simulation, and Figure 4 shows the resulting distribution of average daily doses of chlorpyrifos as determined by the Monte Carlo simulation. Common practice in exposure and risk assessment is to characterize the 50th percentile as a "typical" exposure and the 95th percentile as the "reasonable maximum" exposure.[4] The distributional analysis for these calculated doses

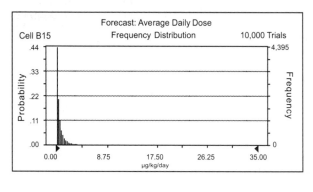

Figure 4 Monte Carlo simulation: distribution of average daily dose.

Table 5 Average Daily Dose of
Chlorpyrifos Estimated from
Monte Carlo Simulation for
Greenhouse Worker

Percentile (%)	Average daily dose (μg/kg/day)
0	0.00
5	0.02
25	0.11
50	0.29
75	0.67
95	2.11
100	31.96

is presented in Table 5. Because this frequency distribution characterizes potential exposures to a given population, the conservatism of the risk assessment should be based on the percentile used. The 50th and 95th percentile exposures for greenhouse workers who apply chlorpyrifos in a microencapsulated formulation are 0.29 and 2.11 μg/kg bw/day, respectively.

Summary and conclusions

For comparison purposes, the results of the various analyses and risk estimates for the mixers/loaders/applicators in greenhouses are summarized in Table 6. All of these analyses present the risk assessor with some information about the possible hazards to workers who are exposed to

Table 6 Comparison of Exposure and Risks for
Greenhouse Mixers/Loaders/Applicators

	Exposure dose (μg/kg/day)	Risk assessment (vs. 100 μg/kg/day)
Passive dosimetry		
Arithmetic mean	3.1	MOE = 32
Geometric mean	0.7	MOE = 137
Monte Carlo		
50th percentile	0.29	MOE = 345
95th percentile	2.11	MOE = 47
Biomonitoring		
Arithmetic mean	1.0	MOE = 100
Geometric mean	0.5	MOE = 200
Statistical analysis	All values	$p < 0.0001$ (<1 in 10,000)

Note: MOE = margin of exposure.

pesticides. Use of the arithmetic mean is clearly the most conservative approach and probably does not represent "typical" exposure unless the data exhibit low variability. The Monte Carlo simulation probably provides the best estimate of central tendency (50th percentile) because it considers the entire range of possible doses to workers rather than just those that were included in the study. The best approach may be the statistical probability analysis, because this method gives the risk assessor an estimate of how often individuals may be expected to approach a specific dose value. This approach is also more easily accomplished than the Monte Carlo simulation, which can require the acquisition of substantial amounts of data other than the exposure data obtained in a field study as well as much larger computational time. In any case, whether the exposure is acceptable or not should be based on the NOEL rather than the reference dose (R_fD).

References

1. Honeycutt, R.C., Day, Jr., E.W., Shurdut, B.A., and Vaccaro, J.R., Use of simultaneous biological monitoring and dermal dosimetry techniques to determine the exposure of chlorpyrifos to applicators and re-entry workers, in *Worker Exposure to Agrochemicals: Methods for Monitoring and Assessment*, Honeycutt, R.C. and Day, E.W., Jr., Eds., Lewis Publishers, Boca Raton, FL, 2000, chap. 2.
2. Nolan, R.J., Rick, D.L., Freshour, N.L., and Saunders, J.H., Chlorpyrifos: pharmacokinetics in human volunteers, *Toxicology and Applied Pharmacology*, 73, 8–15, 1984.
3. Shamoo, D.A., Johnson, T.R., Trim, S.C., Little, D.E., Linn, W.S., and Hackney, J.D., Activity patterns of a panel of outdoor workers exposed to oxidant pollution, *Journal of Exposure Analysis and Environmental Epidemiology*, 1(4), 423–438, 1991.
4. U.S. EPA guidelines for exposure assessment, *Federal Register*, 57, 22888, May 29, 1992.
5. Shurdut, B., Assessment of Exposures and Risks to Greenhouse Workers Applying a Chlorpyrifos-Based Insecticide to Ornamentals Using Conventional and Probabilistic Approaches, paper presented at the International Union of Pure and Applied Chemistry, 1994.

chapter four

Human dose comparisons utilizing biomonitoring and passive monitoring of an exposure environment following surface treatment with an insecticide

J.R. Vaccaro, R.J. Nolan, and M.J. Bartels

Contents

Introduction

Major questions that arise whenever a pesticide exposure evaluation is completed are how good are the data and how close to the real answer have we gotten? For most commercially sold insecticides, there are no appreciable pharmacokinetic data in human systems, although some data normally exist for animal models. Because such pharmacokinetic data do not exist for most active insecticides, passive dosimetry measurements must be used to estimate the exposure and eventually dose. Once such passive dosimetry data exist, certain assumptions must be made to arrive at an estimate of dose.

It is the purpose of this paper to share the results of two studies in which both biomonitoring and passive dosimetry were conducted simultaneously. Comparisons were then made to see how much agreement existed between the two approaches in the assessment of dose in a single study. It is generally accepted that the biomonitoring results are closer to the actual value, as biomonitoring yields a measure of the actual amount of test material absorbed into the body or a measure of the quantity of metabolite. The objective of these studies was to assess how the results of using passive dosimetry methodology compared with the results of using urinary biomonitoring only.

Materials and methods

This discussion pertains to our comparison of the use of both biomonitoring and passive dosimetry techniques to estimate the exposure of a group of human volunteers to chlorpyrifos upon re-entry to a lawn previously treated with the insecticide. Our studies included two types of applications to two different lawns. The first application was conducted by a professional lawn-care specialist applying a liquid formulation. The second application involved the spreading of a granular formulation onto a grassy surface by a non-professional applicator. Both applications were followed by human activity on that surface, and both passive monitoring and biomonitoring were conducted in both studies (discussion of the liquid application will be presented in some detail). In both cases, the test substance was chlorpyrifos, an insecticide registered for use on lawns. Chlorpyrifos is an organophosphate, a white solid at room temperature with a melting point of approximately 41°C and a vapor pressure of 1.8×10^{-5} mmHg at 25°C. Chlorpyrifos has very low water solubility but high solubility in acetone and hydrocarbon solvents.

Application was made in each study to a 40,000 square foot (3716 m^2) grassy lawn. The liquid application was made with a 0.3% aqueous mixture at a rate of 4 lb active per acre. This is the highest labeled rate and is normally used for control of grubs. The granular application was completed with a cyclone spreader, similar to spreaders that may be purchased by homeowners at the local hardware store. The free-flowing granule was 0.5% active on a substrate of ground corn cobs. The application rate for the granular deposition was approximately 2 lb per acre, the maximum labeled rate and one half the rate used for the liquid application. Deposition uniformity was measured by placing numerous sample coupons on the grass prior to application. These coupons were then analyzed for chlorpyrifos to determine application uniformity.

To monitor the absorbed chlorpyrifos doses in human volunteers, urine was collected before and following a 4-hr activity period on the treated grass surface(re-entry). The urine was analyzed for 3,5,6-trichloropyridinol (3,5,6-TCP), the urinary metabolite of chlorpyrifos, and creatinine, which was determined to verify completeness of urine collection by each volunteer.

Doses of chlorpyrifos in human volunteers were also estimated using physical measurements. Air sampling was conducted in order to estimate the inhalation dose to each volunteer. Dislodgeable residues were also measured throughout the study to estimate the dermal contribution to total dose. Finally, hand rinses were conducted on each volunteer immediately following the 4-hr activity period to assess the potential contribution to total dose from hand exposure and to estimate an oral dose to a crawling child.

Air sampling

Disposable cassettes with filters and glass absorbent tubes containing Chromosorb 102 chromatographic packing (66-mg front section, 33-mg back section; SKC, Inc., Eighty Four, PA) were used, in tandem, to trap chlorpyrifos aerosol and vapor, respectively. In the liquid turf study, air was drawn through a GN-4 mixed cellulose ester filter (37-mm diameter, 0.8-mm pore size; Gelman Sciences, Ann Arbor, MI), housed in a cassette, and glass tubes at the rate of approximately 1 L per minute using portable battery-operated vacuum pumps. This sampling train was attached to the battery-operated pump using flexible tubing. The cassette was used in the closed face position with only the small hole unobstructed to capture any airborne particulate. A sampling time of approximately 1 hr was used which allowed for the construction of a dissipation profile curve. To determine inhalation doses, 4-hr time-weighted averages (TWA) were determined during the 4-hr activity period. Determination of the flow rate was made using calibrated rotometers. Pump flow rates were measured both before and after sampling. Sampling occurred using an "F"-shaped metal pole. The top rung was approximately 5 feet (1.5 m) from the ground and the lower rung approximately 15 inches (38 cm) from the ground. This represented the approximate breathing zones of an adult standing and weeding or a child crawling or sitting. TWAs were determined at each height for the 4-hr activity period. The sampling took place at the approximate center of the test field.

Dislodgeable residues

The objective of determining dislodgeable residues was to ultimately approximate the dermal exposure and dose contribution from the dermal route for each of the nine adult volunteers. Dislodgeable residues were determined by use of the Dow drag system. The system utilized a lead weight over a 3" × 3" × 3/4" plywood block. The weight could be varied to represent the type of volunteer being evaluated (i.e., man, woman, or child). Affixed to the bottom and side of the plywood block was a 4" × 4" piece of tightly woven denim; the value of this system is that uniform pressure is applied to the test surface. Two weights were used in the liquid turf study, one to approximate the pressure exerted by a crawling 1-year-old child weighing 10 kg, and one to approximate the pressure exerted by a walking adult female weighing approximately 60 kg. The entire drag system containing the denim coupons was dragged over the treated surface. The drags were taken in 48" lanes such that, when the analysis was conducted, all data were given in micrograms of material dislodged per square foot. These lanes were established using a wooden template. Following the drying period of 4 hr (liquid application), five drags were taken with the small weight and five with the heavier weight at varying time intervals. Similar testing occurred at 8, 12, 24, and 48 hr post-application. All drag coupons were stored frozen until analyzed.

Hand rinses

At the termination of the activity period, each participant's hands were held over a bowl and doused with 250 mL of a dilute dioctyl sodium sulfosuccinate (anionic surfactant) mixture. This soap wash was followed by a 250-mL rinse with deionized water. The soap and water fractions were stored together in the same container. Fifteen grams of sodium chloride were added to the container to facilitate phase separation. The chlorpyrifos was partitioned with 200 mL of ethyl acetate, which was also used to rinse the bowl. The ethyl acetate extract was later analyzed for chlorpyrifos content. The amount of test substance removed was used to assess adult hand exposures and dose and also to assess the theoretical amount of test substance removed when children put their hands in their mouths.

Human activity

In the liquid study, 4 hr were allowed for drying before the volunteers were allowed to commence exercises on the treated surface. In the granular application, no time was needed for drying; therefore, activity began shortly after completion of the application. In the liquid study, human activity was conducted on the west half of the 40,000 ft² field. The five activities consisted of touch football, which was played for a total of 60 minutes; sunbathing, for a total of 30 minutes; frisbee, for a total of 60 minutes; weeding on hands and knees, for a total of 30 minutes; and picnicking (sitting on a blanket), for a total of 60 minutes. The activities were conducted for 15-minute intervals throughout the 4-hr activity period. The touch football represented a child walking/running, the sunbathing represented a child sleeping, the frisbee activity represented a child walking, the weeding represented a child crawling, and the picnicking represented a child sitting and playing on the turf. The sunbathing was conducted on a towel, which is quite typical for sunbathing, but the activity was conducted under an awning to reduce the amount of sun exposure. The picnicking was also conducted under an awning for the same reason. The total elapsed time for all the activity periods combined was 240-minutes. Each participant wore only running clothes, which consisted of a t-shirt and shorts. Running shoes were worn only during the touch football portion to prevent any foot injuries. Each volunteer was instructed not to shower until 4 hr after completion of the activity period in an effort to allow for more complete absorption from the skin. The group of volunteers consisted of both males and females.

Volunteers

The volunteers were obtained by placing notices on bulletin boards and/or in mail boxes within the Midland location of The Dow Chemical Company. Participation in this study was limited to adults 21 to 55 years of age who were salaried employees of The Dow Chemical Company or DowElanco, or were part-time employees of either company, and in good general health as determined by medical history and physician's examination. Each participant was examined by a physician prior to being admitted to the study. As part of these examinations, the physician reviewed each candidate's medical history, recent exposure to pesticides, and results of recent laboratory tests (electrocardiogram, CBC, urinalysis, and the following clinical chemistries: alkaline phosphatase, SGOT, SGPT, LDH, total bilirubin, BUN, total protein, globulin, albumin, and serum creatinine). All females of childbearing age were tested to verify that they were not pregnant. The examining physician advised each volunteer of the results and meaning of any abnormal laboratory test. The purpose of these examinations was to document the health status of the participants and to identify and exclude individuals from participating in the study who were taking chronic medication; who had liver, kidney or cardiovascular disease; or who had recent exposure to pesticides for which 3,5,6-TCP is a possible metabolite or degradation product. Examples of these pesticides would be chlorpyrifos, triclopyr, and methyl chlorpyrifos. The examining physician notified the participants in writing as to whether an individual was allowed to participate in this study and whether or not any changes in the health status were found in the post-cholinesterase testing. Each volunteer was fully advised as to the purpose of the study and of the potential hazards and inconveniences associated with participation in this study. Prior to being admitted to the study, each volunteer was given a copy of an informed consent form to read and sign after proper discussion of its contents. Some monetary inducement was offered to the volunteers for their participation. The participants were instructed to abstain from all medication including aspirin and non-prescription drugs during the period when urine specimens were being collected.

Collection of biological specimens

Blood

Blood specimens of approximately 5 mL were collected on two separate days during the week preceding the study. Additional blood specimens of approximately 5 mL each were collected approximately 24 and 48 hr after the start of the study. These blood specimens were drawn and assayed for plasma cholinesterase activity by personnel from the Michigan Division Medical Department of The Dow Chemical Company.

Urine

Each participant was asked to collect all urine voided the day prior to the start of the study and on the first 5 days following the start of the study as well as the day of the study. Urine voided on each of these days was collected as two specimens, each representing all urine voided during an interval of approximately 12 hr. The first collection each day represented the urine voided between the first voiding in the morning and approximately 7 p.m. in the evening. The second collection period started with the end of the first collection and ended with the first voiding on the next morning. To collect these specimens, the volunteers were instructed to empty their bladder at the start and end of each 12-hour collection interval, and to record the time the collection interval actually began and ended on the tag affixed to the container. Urine voided at the start of the first pre-study collection period was discarded. Urine voided at the start of the other intervals was added to the container for the previous collection. These specimens were collected at ambient temperatures in tared 4-L amber-tinted Polypac sample containers.

Upon receipt by the analyst, the urine specimens were weighed and the volume of each specimen calculated from its weight assuming a specific gravity of 1.00 for the urine. Aliquots (~30 mL) were then removed, transferred to 40-mL amber VOC vials, and stored frozen until analyzed for creatinine and 3,5,6-TCP. The vials containing the aliquots of urine were labeled with the same information contained on the tag attached to the collection container. The information on the tag and the full and empty weight of each urine collection container were recorded in a sample log. Urinary creatinine was measured using a modification of the method described by Fabiny and Ertinghausen,[1] which is based on the Jaffe reaction. Urinary 3,5,6-TCP was extracted and quantified using the method described by Nolan et al.[2] To document that 3,5,6-TCP was not lost during the storage or analysis, aliquots of the reference urine specimens were fortified with a known amount of 3,5,6-TCP. These fortified urine samples were stored and analyzed for 3,5,6-TCP with the other urine specimens. Following the activity period on the treated surface, all subjects were instructed to shower no sooner than 4 hr later. This added consistency to each subject's subsequent activities during which urine specimens continued to be collected.

Analysis of urine.[3] The amount of creatinine and TCP in each urine collection was calculated from the volume of the urine specimen and the concentration of each in that urine specimen. The urine was analyzed for TCP by acidifying 5 mL with 0.5 mL of HCl, followed by heating at 80°C for 2 hr. Upon cooling, the samples were extracted with ethyl ether (1 × 5 mL). The ether was stripped in a stream of nitrogen to dryness. The residues were taken up in 1 mL of *o*-xylene and derivatized with 100 µL of *N-t*-butyldimethylsilyl-*N*-methyltrifluroacetamide (MTBSTFA) at 60°C for 1 hr. The TCP derivative was then analyzed by mass spectral analysis and gas chromatography with a Finnigan TSQ-70 GC/MS under the following conditions:

- 30-m J&W DB-5 fused silica capillary column (0.25 mm i.d. × 0.25 µm film; J&W Scientific)
- Helium carrier gas, 1 to 2 mL/min (0.2 min splitless injection)
- Gas chromatographic oven temperature program: 220°C for 6.5 min, 40°C/min to 260°C; injection and capillary transfer line at 280°C.
- Representative mass spectral conditions (negative chemical ionization): ion source temperature, 150°C; ionizing current, 0.20 mamp; electron energy, 70 eV; methane reagent gas (source pressure 0.5 to 1 torr).

Quantitative data for the 3,5,6-TCP and $^{13}C_2$-3,5,6-TCP were obtained by selected ion monitoring of the dichloropyridinol fragment ions (m/z 161 to 165; 0.1 sec/scan). The amount of TCP found in the urine was used to calculate the amount of chlorpyrifos represented by that amount of TCP based on molecular weight differences.

Analysis of the urinary data. The amount of creatinine and 3,5,6-TCP in each urine collection was calculated from the volume of the urine specimen and the concentration of each in that urine specimen. The amount of creatinine excreted per day was compared across days for each volunteer and to standard literature values for creatinine excretion (i.e., mean 1.8 g/24 hr; 95% range, 1.1 to 2.5 g/24 hr). The urine collection was considered to be complete if the amount of creatinine was consistent with the amount of creatinine in the other urine specimens provided by that individual and within the literature range for normal creatinine excretion.

Two methods were used to calculate the amount of chlorpyrifos absorbed. First, the amount of chlorpyrifos absorbed was estimated by dividing the amount of 3,5,6-TCP in the urine collected 0 to 144 hr after the start of the study by 0.343. This factor represented the product of the ratio of the molecular weight of 3,5,6-TCP and chlorpyrifos (i.e., $198/350.6 \approx 0.56$) and the fraction of the dermally absorbed chlorpyrifos expected to be excreted in the urine in 120 hr (0.613). Second, the amount of chlorpyrifos absorbed was estimated by fitting the interval 3,5,6-TCP excretion data to the one-compartment model used by Nolan et al.[2] to describe the time-course of 3,5,6-TCP in urine of volunteers following application of chlorpyrifos to their forearm. Nolan et al.[2] found that an average of 71% of the orally administered chlorpyrifos was excreted in the urine as 3,5,6-TCP, and that the time-course of 3,5,6-TCP in the urine following dermal administration was well described using an absorption and elimination rate constant of 0.0308/hr and 0.0258/hr, respectively. Optimized estimates for the amount of chlorpyrifos absorbed were based on the maximum likelihood function, obtained using SimuSolv®, a computer modeling program that contains integration, optimization, and graphical routines.

Results

Doses for each volunteer were estimated by two methods. The first approach was to sum the contributions of the inhalation and dermal routes (whole body and hands). These contributions were estimated using calculations based on the air sampling data, the dislodgeable residue data, and the hand-rinse data.

The second estimate of dose was arrived at through analysis of the collected urine of volunteers for 3,5,6-TCP and creatinine. The second approach is generally considered to yield a more accurate estimate of dose as it measures, indirectly, the amount of test substance absorbed from all routes of exposure. The following estimation of dose is for the liquid turf evaluation; only a summarization of data will be given for the granular turf evaluation.

Dose based on physical measurements

Inhalation

Each of the activities conducted on the lawn was categorized as rest, light activity, or heavy activity, as the type of activity influenced the breathing rate. Frisbee and touch football were classified as heavy activity (120 min); weeding was classified as light activity (30 min); and the combination of picnicking and sunbathing was classified as rest (90 min). The total duration of the activity period was 4 hr.

Four-hour time-weighted averages were determined during the activity period. The measured 4-hr TWA at the 15-inch (38-cm) height was 15.2 mg/m^3, and the TWA at the 5-foot (1.5-m) height was 5.08 mg/m^3. Breathing rates at rest and during light and heavy activity for females are listed as 6 L/min, 19 L/min, and 25 L/min, respectively.[4] The breathing rates at rest and during light and heavy activity for the male are listed as 7.5 L/min, 20

L/min, and 43 L/min, respectively.[4] Therefore, the amount of chlorpyrifos absorbed via the inhalation route for the female and male was calculated as:

Females
Light activity: 30 min × 19 L/min × 0.001 m³/L × 15.2 µg/m³ = 8.66 µg
Rest: 90 min × 6 L/min × 0.001 m³/L × 15.2 µg/m³ = 8.21 µg
Heavy: 120 min × 25 L/min × 0.001 m³/L × 5.08 µg/m³ = 15.24 µg
Total = 32.1 µg

Males
Light activity: 30 min × 20 L/min × 0.001 m³/L × 15.2 mg/m³ = 9.12 µg
Rest: 90 min × 7.5 L/min × 0.001 m³/L × 15.2 µg/m³ = 10.26 µg
Heavy: 120 min × 43 L/min × 0.001 m³/L × 5.08 µg/m³ = 26.21 µg
Total = 45.6 µg

By dividing the amount of absorbed chlorpyrifos by the individual body weights, the inhalation dose for the adult volunteers was estimated. Table 1 gives the individual inhalation doses. The mean inhalation dose for the adult volunteers was 0.59 mg/kg.

Table 1 Liquid Turf Study: Adult Inhalation Dose of Chlorpyrifos Based on Air Monitoring Data

Volunteer	Weight (kg)	Chlorpyrifos absorbed (µg)	Dose (µg/kg)
1	59	32.1	0.54
2	88.5	45.6	0.52
3	49.9	32.1	0.64
4	77.2	45.6	0.59
5	74.5	45.6	0.61
6	78.1	45.6	0.58
7	80.4	45.6	0.57
8	49.9	32.1	0.64
Mean			0.59

Estimated dermal dose
The adult dermal chlorpyrifos dose was determined using:

$$\frac{DR \times 1000 \text{ ft}^2 \times 0.01}{WT \text{ (kg)}} = \text{dose (µg/kg)}$$

where:

DR = Mean dislodgeable residue during the activity period (average of T4 and T8 values)
1000 ft² = Surface contact area (estimated approximate area contacted)
0.01 = 1% dermal absorption factor (EPA modification of Nolan et al.[2])

The mean dislodgeable residue was determined by taking the average of the T4, high-pressure dislodgeable residue value (27.8 µg/ft²) and the T8 high-pressure dislodgeable residue value (16.3 µg/ft²), or 22.05 µg/ft². Table 2 summarizes the dermal contributions to total chlorpyrifos dose. The average dermal dose was 3.28 µg/kg.

Table 2 Liquid Turf Study: Adult Dermal Dose of Chlorpyrifos Based on Dislodgeable Residue Data

Volunteer	Weight (kg)	Average dislodgeable residue (µg/ft²)[a]	Dose (µg/kg)
1	59	22.05	3.71
2	88.5	22.05	2.48
3	49.9	22.05	4.39
4	77.2	22.05	2.84
5	74.5	22.05	2.94
6	78.1	22.05	2.81
7	80.4	22.05	2.71
8	49.9	22.05	4.39
Mean			3.28

[a] Average of T4 and T8 dislodgeable residue values.

Estimation of dermal (hand) dose

The adult dermal dose resulting from contact of grass with hands was determined through a hand wash of each volunteer at the termination of the activity period. The dermal hand dose was estimated using:

$$\frac{AR\ (\mu g) \times 0.01}{Wt.\ (kg)} = dose\ (\mu g/kg)$$

where:

AR = Amount of chlorpyrifos removed in the hand rinse
0.01 = 1% absorption factor (EPA modification of Nolan et al.[2])
Wt. = Body weight

Table 3 shows the estimated chlorpyrifos dose stemming from absorption of chlorpyrifos from the hands. The approximate average amount of chlorpyrifos absorbed was calculated to be 0.007 µg/kg.

Table 4 summarizes the results of using physical techniques to estimate total chlorpyrifos doses of adults following activity on treated grass. Total doses ranged from 3.03 µg/kg to 5.04 µg/kg (mean, 3.88 µg/kg). Approximately 85% of the chlorpyrifos dose came from the whole body dermal route. About 15% came from the inhalation route. Hand exposure was insignificant.

Table 3 Liquid Turf Study: Adult Dermal (Hands) Dose
of Chlorpyrifos Based on the Handwash Analysis

Volunteer	Weight (kg)	Amount removed (μg)	Dose (μg/kg)
1	59	72	0.012
2	88.5	122	0.013
3	49.9	18.4	0.004
4	77.2	40.2	0.005
5	74.5	54.2	0.007
6	78.1	37.6	0.005
7	80.4	14.5	0.002
8	49.9	17.6	0.004
Mean			0.007

Table 4 Liquid Turf Study: Total Adult Chlorpyrifos
Dose Based on Physical Measurements

Volunteer	Inhalation dose (μg/kg)	Dermal dose (μg/kg)	Dermal (hands) dose (μg/kg)	Total dose (μg/kg)
1	0.54	3.71	0.012	4.25
2	0.52	2.48	0.013	3.03
3	0.64	4.39	0.004	5.04
4	0.59	2.84	0.005	3.5
5	0.61	2.94	0.007	3.62
6	0.58	2.81	0.005	3.39
7	0.57	2.71	0.002	3.27
8	0.64	4.39	0.004	5.04
Mean	0.59	3.28	0.007	3.88

Estimation of dose from urinary metabolite analysis

From the analysis of volunteer's urine for 3,5,6-TCP, the amount of absorbed
chlorpyrifos was determined for each individual volunteer. Table 5 summarizes the calculated chlorpyrifos dose based on analysis of urine samples.
The average chlorpyrifos dose was estimated to be 7.07 μg/kg, 182% of the
dose estimated using physical techniques.

The granular turf study was handled in a manner very similar to the
liquid turf study discussed above. As might be expected, the estimated doses
of chlorpyrifos for nine volunteers were considerably lower than the estimated doses in the liquid study. The volunteers' chlorpyrifos doses ranged
from 0.31 μg/kg to 0.44 μg/kg using physical measurements; average 0.36
μg/kg. Utilizing analysis of urine, the average dose was estimated to be 1.3
μg/kg (Table 6).

Table 5 Liquid Turf Study: Estimated Doses
of Chlorpyrifos Based on Biomonitoring
and Physical Measurements

Volunteer	Physical dose (μg/kg)	Biomonitoring dose (μg/kg)
1	4.25	2.5
2	3.03	10
3	5.04	10
4	3.5	7.6
5	3.62	5.2
6	3.39	5.5
7	3.27	8.6
8	5.04	6.7
Mean	3.88	7.07

Table 6 Dose Comparisons in the Granular Turf Study

Volunteer	Weight (kg)	Inhalation dose (μg/kg)	Dermal dose (μg/kg)	Dermal (hands) dose (μg/kg)	Passive total dose (μg/kg)	Bio-monitoring total dose (μg/kg)
G	79.45	0.019	0.31	0.002	0.33	0.54
H	73.55	0.02	0.33	0.001	0.34	2.48
I	79.45	0.019	0.31	0.001	0.32	0.37
J	81.72	0.018	0.3	0.001	0.31	—[b]
K	74.91	0.02	0.33	0.001	0.33	0.77
L	63.56	0.019	0.39	ND[a]	0.39	0.9
M	72.64	0.017	0.34	ND	0.34	0.54
N	56.75	0.022	0.43	ND	0.44	1.43
O	57.2	0.022	0.43	ND	0.43	3.24
Mean					0.36	1.30

[a] ND = non-detected.

[b] Unable to calculate due to unusually high creatinine levels.

Discussion

An attempt was made to develop physical sampling methodology for the estimation of absorbed doses of human volunteers conducting activities on a lawn previously treated with an organophosphate insecticide. Estimates of inhalation doses and dermal doses were made using available air-sampling data and dislodgeable-residue data. In addition, analysis of volunteers' urine for the urinary metabolite of chlorpyrifos yielded estimations of volunteers' chlorpyrifos doses. Comparison of the results of the two methods indicated that reasonably good agreement was seen. In the liquid study, the average estimated chlorpyrifos dose for all the volunteers, using physical data, was 3.88 µg/kg; using biomonitoring data based on the analysis of the urine, it was 7.07 µg/kg. In the granular turf study, the estimated doses from the physical and biomonitoring data were 0.36 µg/kg and 1.3 µg/kg, respectively. Because the largest contribution of exposure (and dose) stems from dermal contact with the grass, the contact surface area chosen is critical for there to be good agreement of the methods. For example, in the liquid study, if we had approximated the contact area closer to 2200 ft^2, the two values for absorbed chlorpyrifos from the physical and biomonitoring approach would have been nearly identical.

In this physical methodology, hand-rinsing the volunteers immediately following the activity period removed some test substance from the hands; if left on the hands, this substance would have had the opportunity to be absorbed and therefore would have increased the biomonitoring values. In addition, we do not currently know the efficiency of the hand rinse. It is generally thought that the efficiency of the hand rinse using anionic surfactants to remove chlorpyrifos is no better than 50%,[5] a value to be determined in future research.

Because the majority of test materials have no pharmacokinetic data generated in human volunteers, it is important to develop reliable physical study methodology, the results of which would approximate data generated in human volunteers using urinary metabolite analysis. Although agreement in this study was reasonably good, it would be imprudent to say without further studies that all test substances would give such agreement between methodologies. It would appear from the data generated in this report that the physical study design we used sufficiently approximates the estimation of dose generated from the analysis of urine for the urinary metabolites of chlorpyrifos.

References

1. Fabiny, D.L. and Ertinghausen, E., Automated reaction rates method for determination of serum creatinine with the certifichem, *Clinical Chemistry*, 17, 696–700, 1971.
2. Nolan R.J. et al, Chlorpyrifos: pharmacokinetics in human volunteers, *Toxicology and Applied Pharmacology*, 73, 8–15, 1984.
3. Bartels, M.J. and Kastl, P.E., *Journal of Chromatography*, 574, 69–74, 1992.
4. Snyder, W.S. et al., *Report of the Task Group on Reference Man*, International Commission on Radiological Protection, #23, Pergamon Press, Oxford (adopted by the commission in October 1974).
5. Vaccaro, J.R. et al., The use of unique study design to estimate exposure of adults and children to surface and airborne chemicals, in Tichenor, B.A., Ed., *Characterizing Sources of Indoor Air Pollution and Related Sink Effects*, ASTM STP1287, American Society of Testing Materials, 1996, pp. 166–183.

chapter five

Determination of the efficiency for pesticide exposure reduction with protective clothing: a field study using biological monitoring

D.H. Brouwer, S.A.F. De Vreede,
W.J.A. Meuling, and J.J. van Hemmen

Contents

Abstract

A study was conducted to evaluate the effectiveness of protective clothing on the reduction of dermal exposure to pesticides in field practice. The study was designed as an intervention type of study to afford "within-worker" comparisons. The study was performed in nine commercial greenhouses used for the cultivation of carnations in the Greenhouse district in The Netherlands. The insecticide propoxur was used as a test substance. Exposure of the hands as well as respiratory exposure of nine applicators and 18 harvesters were monitored during work with different clothing scenarios. During trial number I, the workers wore normal work clothing; whereas, in trial number II, additional types of protective clothing (including protective gloves) were worn. In addition, biological monitoring (i.e., monitoring of the excreted metabolite 2-isopropoxyphenol [IPP] in 24-hr urine) was performed to assess propoxur absorption. Skin moisture was monitored at five locations of the body. Prior to the intervention study, a feasibility study was conducted using whole-body monitoring to evaluate the variability and the distribution of dermal exposure.

The major area (approximately 60%) of the potential exposure of the applicators was located on the legs, whereas 50% of the potential exposure of the harvesters was equally distributed to the hands and torso. Total body exposure differed significantly between workers.

Actual exposure of the hands was reduced by approximately 95% and 87% for applicators and harvesters, respectively, wearing gloves. Respiratory exposure did not differ significantly between the two trials and was estimated to contribute approximately 4% to the excreted amount of metabolite in urine. Median IPP excretion was reduced 42% and 38% for applicators and harvesters, respectively. IPP excretion of the harvesters was strongly correlated with the actual exposure of the hands during exposure with normal work clothing ($r = 0.93$) and moderately ($r = 0.5$) during exposure with protective clothing. For applicators, low associations of IPP excretions and hand exposure were observed. Skin moisture at the observed locations of the body was significantly higher during the trials with protective clothing. Addition of skin moisture parameters to the linear regression model, as well as the addition of the differences of hand exposure between the two trials (as independent variables), increased the variation of the dependent variable (the differences between IPP excretion of both trials). The results show a substantial reduction of actual exposure by wearing protective clothing; however, the reduction was not fully reflected by a similar reduction of the absorbed amount of propoxur. The data suggest an increased uptake of the remaining actual dermal exposure under the conditions of protective clothing.

In conclusion, based on biological monitoring data, the overall effectiveness of protective clothing is limited. Overestimation of the reduction of exposure is possible when data on the reduction of external exposure are utilized.

Introduction

Dermal exposure is considered to be the primary route of exposure for non-volatile or semi-volatile pesticides in agriculture and can constitute a significant occupational health hazard. Recognition of this hazard has prompted the use of closed loading systems and protective clothing to minimize chemical contact with the skin. While occupational hygienists generally consider personal protective equipment to be a control strategy of last resort (Brouwer and van Hemmen, 1994), the use of protective clothing has become commonplace in the agricultural workplace. Current knowledge of the efficiency of protective clothing performance in agriculture field practice is very limited, although research activity in this field has increased substantially in recent years (Methmer and Fenske, 1994). Field studies have been hampered by lack of an adequate methodology for characterizing the performance of protective clothing under realistic conditions.

The evaluation of protective clothing is complicated by several factors. First, exposure can occur over the entire body, requiring the simultaneous study of several types of garments such as gloves and coveralls. Second, the chemical agent may be found in different physical forms: powder or liquid concentrate during mixing, aerosol spray during application, and particulate transfer originated from pesticide foliar residue. Third, exposure to protected regions of the body may occur through at least three pathways: (1) *penetration* (i.e., movement of the agent through fabric due to porosity; (2) *permeation* (i.e., diffusion of liquid following wetting of fabric); and 3) *direct deposition* (i.e., entrance of the agent through openings in the garment). Finally, normal body movements may influence the movement of the agent through fabric or its deposition beneath clothing. A comprehensive technique for evaluating dermal exposure under field conditions must be available to address these many variables.

The ability of fabrics and garments to reduce pesticide exposure has been tested mostly in trials of the transmission of the pesticide through the material both in laboratory (Crowse et al., 1990; Freed et al., 1980; Leonas, 1991; Oakland et al., 1992a; Raheel, 1991) and to a minor extent in field trials (Nigg et al., 1991, 1993; Oakland et al.,1992b). The reduction of external exposure as a result of protective clothing has been evaluated in field studies (Chester et al.,1990a; De Vreede et al., 1994; Dubelman et al., 1982; Fenske, 1988; Fenske et al., 1987; Methner and Fenske, 1994; Nigg et al., 1986, 1990).

Evaluation of the efficiency of exposure reduction in field practice based on results of biological monitoring has the advantage of including all exposure pathways. However, a relatively small number of field studies that include biological monitoring has been published in open literature. Cowell et al. (1987), Maddy et al. (1989), and Ojanen et al. (1992) conducted biological monitoring measurements but did not use the results to draw conclusions as to the effectiveness of protective measures. Lander and Hinke (1992) and Wicker et al. (1979) used the results of biological effect monitoring (i.e., cholinesterase inhibition) for their conclusions about the reduction of worker exposure due to protective measures. Chester et al. (1990b), Davies et al. (1982), and Nigg and Stamper (1983) used the decrease in excretion of metabolites of pesticides in urine to show significant reduction of worker pesticide absorption due to protective clothing. Ojanen et al. (1992) used biological monitoring in addition to field tests to evaluate the external exposure reduction of phenoxy herbicide exposure, and Aprea et al. (1994) evaluated the influence of protective measures on the excretion of alkylphoshates during manual operations with treated plants. In a non-pesticide exposure scenario, van Rooij et al. (1993) evaluated the reduction of skin contamination by protective clothing on the internal dose of creosote workers exposed to PAHs.

In addition, the use of biological monitoring has the advantage that skin penetration under particular conditions of protective clothing is included as well in the approach. The results of a dose-excretion study of propoxur by Meuling et al. (1991) using volunteers indicate a significant increase of the dermal uptake of the compound under conditions of occlusion, where there is increased blood flow, skin temperature, and skin moisture.

In order to determine the significance of the reduction of dermal exposure by protective clothing in field practice for pesticide workers in greenhouses, an intervention study was conducted. In this study, protective clothing was used that theoretically would perform sufficient reduction of exposure and had been shown to result in minimum thermal discomfort as well as minimum obstruction of task performance (van Hemmen et al., 1994). The effect of reducing propoxur deposition on skin of both pesticide applicators and re-entry workers for greenhouse crops was studied using biological monitoring. Other objectives of the study were to explore the influence of protective clothing on skin moisture and its possible effect on dermal permeation of propoxur.

Materials and methods

Study design

The study was conducted in nine commercial greenhouses used for the cultivation of carnations in the Greenhouse district in Zuid, Holland, The Netherlands, during the spring and summer of 1995. It consisted of the following parts.

Part I: feasibility study

In order to evaluate "within-worker" variances of dermal exposure and its distribution over the body, whole-body monitoring during three applications and concomitant re-entry was performed for high-volume (HV) applicators (n = 4) and harvesters of carnations (n = 6).

Part II: intervention study

In this part of the study, the internal dose of propoxur was assessed for HV applicators (n = 9) and harvesters (n = 18) using biological monitoring in two trials. In the first trial, workers wore their normal work clothing, followed by a trial where the same workers wore additional protective clothing. The minimum period between the two trials was 5 days.

Work clothing was defined as clothing that was normally worn by the workers. For the applicators, such clothing included jeans and a long-sleeved shirt; re-entry workers wore jeans and a t-shirt or short-sleeved shirt. Protective clothing consisted of Tyvek coveralls with a hood (DuPont Protech, Polichlo Holland BV; Groningen, The Netherlands) and nitrile rubber gloves (Ansell Edmont Touch & Tuff; Aalst, Belgium), as well as cotton coveralls and stretch-cotton gloves for applicators and harvesters, respectively.

The test substance Undeen®* (active ingredient propoxur, 200 g/L) was applied by hand-held, high-volume spraying equipment at an application rate of a minimum 25 g active ingredient (a.i.) per 1000 m² and a volume rate of approximately 100 L/1000 m².

Dermal exposure assessment

Whole-body monitoring

During Part I of the study, both potential and actual exposure were assessed using the whole-body technique (Chester, 1995). Potential exposure of the hands was measured using pre-washed cotton gloves covering the hands and forearms (stretch-cotton, 200 g/m², surface area [one-sided] 370 cm²; J. van der Wee BV; Riel, The Netherlands). A pair of gloves was used for a maximum period of 1 hour, in order to prevent breakthrough of the pesticide. New gloves were provided after each hour of harvesting, or earlier in case a glove was damaged.

---

* Registered trademark of Bayer AG, Leverkusen, Germany.

Dermal exposure of other body parts was monitored using khaki-colored cotton coveralls (KLM; Haaksbergen, The Netherlands) with an optimal fit. After the sampling period, the coveralls were divided into ten separate pieces each, put into 1- or 5-L polyethylene bottles, and stored at 4 to 7°C until analyzed. The parts of the coveralls represented the body parts — torso (front and back), forearms (left and right), upper arms (left and right), upper legs (left and right), and lower legs (left and right). In addition, a cotton head string was used to sample (front) head exposure. Actual dermal exposure was assessed using long-sleeved cotton t-shirts and cotton underpants, which were divided in a similar way as the overalls.

Actual hand exposure

In Part II, actual exposure of the hands was assessed for applicators after mixing and loading and after application and for harvesters following re-entry. Workers were asked to wash their hands twice with a hypoallergenic soap (Sporex, Kimberly-Clark; Veenendaal, The Netherlands) for 15 seconds followed by rinsing with tap water. A specially designed handwashing device was used that consisted of a tube attached to the water tap of the water supply in the greenhouse, an adjustable flow control set at a flow rate of approximately 1 L/min, a tap, a funnel, and a 5-L polyethylene bottle to collect the rinse water. After the hand washing, the funnel was rinsed and the total volume of the water was assessed by weighing. Immediately after weighing, acetic acid (99% purity) was added to the rinsing samples and the samples were stored until analysis.

Respiratory exposure assessment

Measurements were carried out using an IOM-sampler (IOM, Negretti Automation, England). The sampling head was attached to a constant-flow pump operating at 2 L/min (P2500 Air Sampling Pump, DuPont, U.S.). This set-up was used to estimate the respiratory fraction according to the definition of the American Conference of Governmental Industrial Hygienists (ACGIH, 1985), and contained two glass fiber filters (25 mm, Type A/E, Gelman Sciences, U.S.). One filter operated as a back-up filter to determine break-through of propoxur during sampling. Flows were adjusted before and checked after the sampling period with a pre-calibrated rotameter tube (Rota 1/4-600, Dr. Hennig GmbH, Germany). After air sampling, filters were put into 50-mL polypropylene tubes (Greiner und Söhne GmbH, Germany) separately and stored at –20°C to await chemical analysis.

The total respiratory exposure (RE) during the application or harvesting period was estimated using the following equation:

$$RE_i = V_p \cdot (CBZ_i \cdot t_i) \tag{1}$$

where RE_i = respiratory exposure (µg) during task i (application or harvesting); V_p = pulmonary ventilation during task i (1.9 and 1.25 m³/hr for applicators and harvesters, respectively); CBZ_i = concentration in the breathing zone ($\mu g/m^3$), t_i = duration of task i (hr) (Brouwer and van Hemmen, 1990).

Biological monitoring

The internal dose of propoxur was measured by assessing the total amount of 2-isopropoxyphenol (IPP) excreted in the urine, collected over a period of 24 hr from the start of exposure, and described in detail in previous studies (Brouwer et al., 1993; Meuling et al., 1991). Volunteer kinetics studies revealed a one-to-one relationship of absorbed propoxur and excreted IPP on a mole basis. Based on the results by Machemer et al. (1982), a pulmonary retention of 40% was used to calculate the relative contribution of the respiratory exposure to the internal exposure. To estimate the contribution of the dermal exposure, the calculated respiratory portion was subtracted from the total amount of IPP excreted in urine.

Dislodgeable foliar residue sampling

During re-entry, stratified whole-leaf sampling was conducted. Twenty-four leaves from the harvesting zone of the crop were collected in a 500-mL polyethylene bottle in duplicate. Leaf samples were stored at 4 to 7°C in the laboratory until analysis. After analysis, the leaf volume was measured by stereometric volumetry using a method described by Sherle (1970). A linear relationship between leaf volume and leaf area was determined for carnation leafs by measuring leaf area (one-sided) with a LI-COR 3100 (LI-Cor, Inc.; Nebraska).

Additional observations and measurements

Skin moisture was assessed in triplicate prior to work and immediately after sampling. Locations of measurement were the palm and back of the hands, forearm, "V" of the neck, and front of the head. Measurements were performed using a Corneometer CM820 (Courage and Khazaka; Köln, Germany). Skin moisture was expressed in arbitrary units (AU) representing the electrical capacity of the skin. Temperature and humidity were measured during the trials using a HMP 31 UT temperature/humidity probe (Vaisala; Helsinki, Finland) connected to a data-logger (1200 series, Grant Instruments, Ltd.; Barrington, U.K.). Tank samples were taken to determine spray liquid concentrations.

Quality assurance

Field spikes with laboratory standards as well as spraying solutions were taken at a wide range of concentrations to determine the stability of the samples during both sampling and storage and possible contamination during the sampling procedure. Field blanks were also taken.

Chemical analyses

Cotton sampling matrices were extracted by the addition of a methanol/water solution (60/40% v/v) and adjusted to pH 4.5 using 99% acetic acid and ultrasonification for 10 min followed by a 30-min extraction at 20 strokes per minute. The extract was analyzed using a high-performance liquid chromatography (HPLC) system equipped with a pump (Model 510, Waters; Etten-Leur, The Netherlands), an injection system (Propmis, Spark-Holland), and C-18 cartridge (Bio-Sil HL 90-5, 150 cm, 4.6 mm i.d.) and detected by a SFM 25 fluorescence spectrometer (Kontron Instruments; Maarsen, The Netherlands). The limit of quantification (LOQ) for filters, cotton fabrics, and hand-rinsing liquid was 20 mg/L at an injection volume of 20 mL, whereas the LOQ for dislodgeable foliar residue sampling was 10 mg/L at an injection volume of 50 mg/L. The recovery was >85%, and the stability of the samples during storage was excellent for at least 10 weeks. The IPP (2-isopropoxyphenol) was determined using gas chromatography and mass-selective detection and has been described in detail before (Leenheers et al., 1992).

Statistical analyses and calculations

The data were statistically analyzed using the SOLO Statistical System (BMDP Statistical Software, Inc.; Los Angeles, CA) for personal computers. Differences among work clothing, protective clothing exposures, and environmental conditions were studied using the Wilcoxon test for matched pairs. The association between internal and external exposure and skin moisture was studied by means of regression analyses (Snedecor and Cochran, 1982). A probability of $p < 0.05$ was considered significant. Reduction of both external exposure of the hands and internal exposure was expressed as 100% – (exposure protective clothing/exposure working clothing) × 100%.

Results

Environmental conditions and applications

During application, temperatures ranged from 17 to 27°C and from 16 to 28°C for the work clothing and protective clothing trials, respectively. During harvesting, temperature ranges were 19 to 26°C and 15 to 25°C for both exposure scenarios.

Relative humidity ranges were 45 to 97% and 44 to 98%, and 54 to 97% and 56 to 92% for application and harvesting in both exposure scenarios, respectively. No significant differences were observed between both temperature and relative humidity for both exposure scenarios.

Applications were performed at an average application rate of 36.8 ± 14.3 g a.i./1000 m² and an average volume rate of 113.5 ± 44.5 L/1000 m² (n = 18). The treated acreage ranged from 1000 to 5100 m², and the period of mixing/loading/application ranged from 12 to 77 min (mean = 35.6 min,

STD = 17.8 min). Mixing and loading activities lasted less than 10% of the total working period; no significant differences were observed between the two exposure scenarios for any of the application variables. Resulting dislodgeable foliar residues ranged from <3 to 84 $\mu g/cm^2$ while harvesting with work clothing, and from <3 to 22 $\mu g/cm^2$ while harvesting with protective clothing; no significant differences were observed between the two exposure scenarios.

Dermal exposure

Variances of exposure

The variances of potential dermal exposure are presented in Table 1. Very large "within-worker" variances of potential exposure of the hands resulted in insignificant differences between workers. For harvesters, a similar result was obtained for potential exposure of the body parts. For both applicators and harvesters, significant "between-worker" differences of total potential exposure were observed.

Table 1 Variances of Potential Exposure[a]

Applicators

	Coefficient of variation (CV) (%)			
	Worker 1	Worker 2	Worker 3	Between workers
Hands	62	82	119	NS
Body	11	55	49	—[b]
Total	22	54	62	—[b]

Harvesters

	Coefficient of variation (CV) (%)						
	Worker 1a	Worker 1b	Worker 2a	Worker 2b	Worker 3a	Worker 3b	Between workers
Hands	55	52	76	56	36	41	—[c]
Body	96	68	67	55	57	69	NS
Total	87	65	70	53	55	59	—[b]

[a] The coefficients of variations are given as results of whole-body assessment of dermal exposure during three applications and concomitant re-entries.

[b] ANOVA $p < 0.05$.

[c] ANOVA $p < 0.01$.

Note: NS = not significant.

Figure 1 shows the distribution of the potential exposure. The major area of potential exposure of the applicators (approximately 60%) was located on the legs, particularly the lower legs. Exposure of the legs of the harvesters was approximately half that of the applicators. For harvesters, approximately 25% of the exposure was located on the torso and another 25% on the hands.

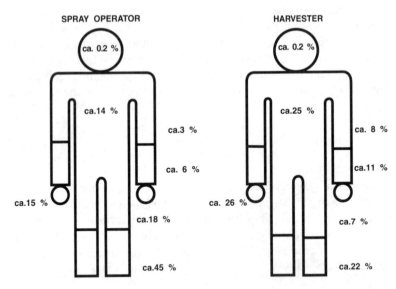

Figure 1 Distribution of the potential dermal exposure of applicators (N = 3, n = 9) to propoxur and of harvesters (N = 6, n = 18).

For applicators, hand exposure was approximately 15% of the total potential dermal exposure.

Actual exposure

Ranges and medians of exposure to the hands are shown in Table 2. Actual exposure was significantly lower when protective gloves were used. The relative contribution of mixing and loading decreased substantially when protective gloves were worn.

Table 2 Ranges (and Medians) of Actual Exposure
of the Hands (in µg Propoxur)

	Applicators (n = 9)			Harvesters (n = 18)
	Mixing/loading	Application	Total	
No gloves	8–5785 (231)[a]	9–416 (122)[b]	31–2390 (348)[c]	7–1523 (164)[d]
Gloves	7–148 (8)[a]	4–47 (8)[b]	11–56 (16)[c]	5–299 (8)[d]

[a–d] Significant differences (Wilcoxon, $p < 0.05$).

Respiratory exposure

Concentrations of propoxur in the breathing zone ranged from 0.6 to 25 (median, 0.7) µg/m^3, and from 0.4 to 29.4 (median, 1.2) µg/m^3 for applicators in exposure scenarios without and with protective clothing, respectively. For harvesters, concentration ranges were 0.1 to 17.8 (median, 0.2) $\mu g/m^3$, and 0.2 to 8.6 g/m^3 (median, 0.5), respectively. Resulting respiratory exposures are given in Table 3. No significant differences were observed between exposure with and without personal protective clothing. The estimated mean contribution of the respiratory exposure to the IPP excretion was about 4%.

Table 3 Respiratory Exposure

	Exposure[a] (μg propoxur)				Contribution to IPP excretion[b] (%)			
	Work clothing		Protective clothing		Work clothing		Protective clothing	
	Mean[c]	Median	Mean[c]	Median	Mean[c]	Median	Mean[c]	Median
Applicators (n = 9)	11.0	0.8	11	0.8	4.7	0.8	4.1	1.2
Range	(0.8[d]–61)		(0.8[d]–76.5)		(0.2–18)		(0.4–16)	
Harvesters (n = 18)	10.7	1.9	6.4	0.5	3.9	1.5	5.4	1.3
Range	(0.5–61.8)		(0.3–41.7)		(0.045–16.6)		(0.1–16.1)	

[a] Respiratory exposure is calculated from the concentration of propoxur in the breathing zone, the duration of the exposure, and the estimated pulmonary ventilation during task performance (i.e., 1.9 and 1.25 m^3/hr for applicators and harvesters, respectively).

[b] The contribution of respiratory exposure is calculated by dividing the theoretically respiratory absorbed nmol of propoxur nmol (0.4 × respiratory exposure) by the total nmol of IPP excreted, assuming that 1 mol propoxur absorbed will result in an excretion of 1 mol IPP.

[c] Arithmetic mean.

[d] 0.8 μg propoxur is derived from the limit of detection for air sampling.

IPP excretion

When normal work clothing was worn, IPP excretion ranged from 128 to 1505 nmol and from 83 to 2189 nmol for applicators and harvesters, respectively. The amount of IPP excreted after working with protective clothing was significantly reduced to 70 to 926 nmol and 16 to 917 nmol for applicators and harvesters, respectively (Table 4).

Skin moisture

Skin moisture ranged from 45 to 147 AU for different parts of the body. Generally, the "V" of the neck and the forearms were the body parts with the highest skin moisture. Average AU values are presented in Table 5. For the applicators, all body parts except the palms of the hand revealed higher skin moisture during the application with protective clothing. Similar results were observed for the harvesters. In this case, only the backs of the hands showed no significant increase in skin moisture during harvesting while wearing protective clothing.

Table 4 Internal Exposure (Dose) of Propoxur Expressed as Excreted Amount of 2-Isopropoxyphenol (nmol IPP)

| | Applicators (n = 9) | | | | Harvesters (n = 18) | | | |
| | IPP_{total} | | IPP_{dermal}[a] | | IPP_{total} | | IPP_{dermal}[a] | |
	Mean	Median	Mean	Median	Mean	Median	Mean	Median
Work clothing	519[b]	207	497	202	570[c]	391	549[d]	390
Range	(128–1505)		(127–1389)		(83–2189)		(82–2188)	
Protective clothing	265[a]	153	244	152	262[c]	206	250[d]	182
Range	(70–926)		(123–789)		(16–917)		(15–916)	

[a] $IPP_{dermal} = IPP_{total} - (40\%$ respiratory exposure [mol propoxur]).
[b-d] Significant differences between groups (Wilcoxon $p < 0.05$).

Table 5A Average (n = 3) Skin Moisture (AU) of Different Body Parts of Applicators (n = 9)

| | Work clothing | | | Protective clothing | | |
	Mean	Median	Range	Mean	Median	Range
Palms of hands	99	102	64–118	102	107	65–120
Backs of hands	96[a]	91	67–121	107[a]	115	64–132
Forearms	100[b]	112	74–114	124[b]	130	89–138
Front of head	98[c]	98	80–123	119[d]	124	83–138
"V" of neck	104[d]	106	76–130	125[d]	129	92–144

Table 5B Average (n = 3) Skin Moisture (AU) of Different Body Parts of Harvesters (n = 18)

| | Work clothing | | | Protective clothing | | |
	Mean	Median	Range	Mean	Median	Range
Palms of hands	97[e]	106	46–118	106[e]	114	46–118
Backs of hands	94	100	45–113	104	114	58–126
Forearms	94[f]	97	53–124	108[f]	119	62–132
Front of head	102[g]	104	71–119	115[g]	115	79–147
"V" of neck	101[h]	107	59–126	118[h]	122	85–140

[a-h] Significant differences between groups (Wilcoxon $p < 0.025$).

Relation between external exposure, skin moisture, and internal dose

The relation between the calculated amount of IPP excreted in urine resulting from dermal exposure and actual exposure of the hands was studied for both work clothing and protective clothing trials using the regression model:

$$IPP_{dermal} = a + b \text{ (hand exposure)} + c \text{ (skin moisture)} \qquad (2)$$

No significant relationship between the IPP excreted and hand exposure was observed for the applicators wearing normal work clothing or for protective clothing ($p = 0.09$ and $p = 0.73$, respectively). Skin moisture variables did not contribute significantly to the explained variation.

For harvesters, hand exposure contributed significantly to the explained variation for trials with normal working clothing ($p < 0.0001$, $R^2 = 0.87$) and with protective clothing ($p = 0.01$, $R^2 = 0.25$). Skin moisture variables slightly increased the variation (R^2 up to 0.32) where the regression model remained significant. However, the influence of the skin moisture variables was not significant.

The differences between the two trials were also studied for the IPP excretion and propoxur contamination of the hands and skin moisture as independent variables. For applicators, this resulted in a significant contribution of hand contamination to the model ($p = 0.03$, $R^2 = 0.4$). Addition of skin moisture variables resulted in considerable increase of the R^2 (up to 0.8), although without a significant contribution of the independent variables (Table 6).

Table 6 Results of Linear Regression Analyses[a]

	$P_{variable}$	R^2_{adj} model	P model
Handwash	0.03	0.40	0.03
Skin moisture:			
Palms of hands	0.29	0.47	0.06
Backs of hands	0.91	0.38	0.10
Forearms	0.16	0.80	0.04
Front of head	0.46	0.66	0.09
"V" of neck	0.20	0.78	0.04

[a] Using the model $\Delta IPP_{dermal} = a + b \, \Delta$hand exp $+ c \, \Delta$skin moisture $+ \varepsilon$, where Δ is the difference between the individual applicator results during exposure with and without protective clothing.

Reduction of dose

Results of the calculated reductions of actual exposure of the hands are plotted in Figure 2. As illustrated, there is a large variability among individual workers. The median reduction of the exposure was 95% for the applicators and 87% for the harvesters. Reduction of the dose showed even larger

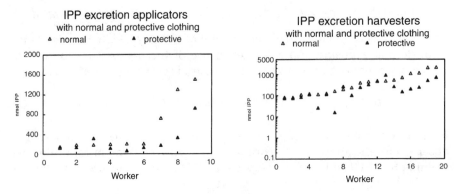

Figure 2 Plots of the excretion of 2-isopropoxyphenol by applicators (n = 9; left side) and harvesters (n = 18; right side) with normal work clothing and additional protective clothing.

variability and negative reduction in a few cases for both applicators and harvesters. The median of reduction of the dose was 42% and 38% for applicators and harvesters, respectively.

Discussion

The design of an intervention type of study demands similar conditions for exposure and environment for the compared situations. In the literature, conclusions on effectiveness of protective clothing are reported based on studies where groups of workers are compared, as well as studies based on "within-worker" comparisons of exposure. In a study by Maddy et al. (1989), strawberry pickers were monitored for three consecutive days during work in the same field. During one of the days, full body monitoring was conducted. On the other two days, workers were randomly selected to wear protective gloves. Based on the relatively low standard deviation of the dislodgeable foliar residue samples, low "between-group" variances were assumed. Chester et al. (1990b) compared the actual exposure of the hands and the urinary excretion of pesticide metabolites within a group of applicators after one day of application. In the study by Aprea et al. (1994) using orchard workers, urinary excretion of two groups of workers (protected and unprotected) were compared. It was assumed that the potential exposures of all workers were the same because they worked in the same orchard and performed the same work, hence small "between-worker" variances were assumed.

Nigg and Stamper (1983) conducted potential exposure and actual exposure measurements during three weekdays over three consecutive weeks for applicators and mixer-loaders, one week using normal work clothes, the second week disposable coveralls in addition to their work clothes, and the third week disposable coveralls and NIOSH-approved respirators. In conjunction, urinary excretion of a metabolite of the active ingredient was monitored. Potential body exposure estimates showed no statistically significant differences at the 95% level for each group between weeks, although differences were observed. It was concluded that weekly cumulative exposure estimates could be used for calculations of protection effectiveness.

The design of a study by Davies et al. (1982) for mixers and applicators was similar to that of Nigg and Stamper (1983). "Between-days" variances of exposure were not given. Mean urinary metabolite concentrations were used to show reduction of internal exposure by protective clothing. The design of the study by van Rooij et al. (1993) was similar to our study (i.e., "within-worker" comparisons of internal exposure). Because no potential dermal exposure was assessed in this study, "within-worker" variances of potential exposure are not known.

In the our study, "within-worker" variances of potential exposure were assessed separately in a feasibility study. Considerable "within-worker" variances of potential exposure were observed; however, "between-worker" variances were even larger, resulting in significant "between-worker" variances for both applicators and harvesters. The large coefficients of variations of exposure of the hands observed within applicators did not contribute substantially to the variances of exposure of the total body as potential exposure of the hands was approximately 15% of the total potential exposure. Assuming a reduction in the actual exposure of 95% by wearing protective gloves and of 90% by wearing coveralls, overall actual dermal exposure would be reduced to 65% and 75% for applicators and harvesters, respectively. The correlation of the effect of intervention (i.e., protective clothing), the resulting internal exposure was assumed to be 0.7. Mean "within-worker" variances of external exposure (expressed as the coefficient of variation, or CV) calculated from the data in Table 1 were 46% and 65% for applicators and harvesters, respectively. To achieve a power of 0.8 with $a = 0.5$, the minimum sample size of a repeated-measures designed study was calculated to be 7 applicators and 11 harvesters.

Other source-strength estimates for exposure (i.e., respiratory exposure data and dislodgeable foliar residue) and environmental conditions (i.e., temperature and relative humidity) did not differ significantly between the trials. Together with the observed "within-worker" variances of exposure, this supports the assumption of similar exposures for both clothing scenarios; therefore, it meets the conditions of an intervention type of study design. Conclusions about the reduction of external exposure by protective clothing in our study can only be given for exposure of the hands (i.e., comparison of actual exposure of the hands with and without gloves). The median efficacy of gloves in field practices (i.e., 95% and 87% for applicators and harvesters, respectively) fit well into the observations of other studies. Nigg et al. (1986) observed an 84% reduction for exposure during mixing/loading and application. Chester et al. (1990b) reported, on average, an 87% reduction in actual hand exposure of mixers/loaders/applicators. Maddy et al. (1989) observed mean actual hand exposures of re-entry workers working with cotton gloves that were approximately 2 to 3% of the exposure of those working without gloves. Aprea et al. (1994) also observed large differences between re-entry workers wearing cotton gloves compared to an individual wearing no gloves; however, the limited sample size precluded definitive statements on the reduction efficacy in this study.

Prior to drawing conclusions about internal exposure and the overall efficacy of protective clothing, the contribution of different pathways should be evaluated; therefore, respiratory uptake should be taken into consideration. Based on the assumption of a pulmonary retention of 40%, the median contribution of respiratory exposure of propoxur to IPP excretion was approximately 4% for both scenarios and fits into the range of 2 to 20% observed in a previous study (Brouwer et al.,1993). This might be an underestimation, as the capture efficiency of the air-sampling device will also result in the capture of inhalable but non-respirable aerosols (Mark and Vincent, 1986). Meuling et al. (1991) showed 100% retention of orally administered propoxur in volunteers, so secondary ingestion of non-respirable particulates will result in a higher absorbance of propoxur than the pulmonary route. In a worst-case scenario, assuming that all aerosol containing propoxur was non-respirable, this would theoretically result in a contribution to IPP excretion by the respiratory route of approximately 10%. Therefore, it can be assumed for the purpose of this study that the true contribution to IIP excretion would be between 4 and 10%. Davies and coworkers (1982) observed a slight nonsignificant reduction of urinary metabolite excretion after the day that workers wore respirators in addition to protective clothing compared to the day when only protective clothing was used. Comparable results were reported by Nigg and Stamper (1983), who observed a nonsignificant average decrease of urinary metabolite excretion of approximately 5% during the clothing modality that included the respirators. Aprea et al. (1994) observed significant different biological monitoring results between workers with and without respirators and concluded that the results suggest that the respiratory system is an important route of absorption of azinphos-methyl and chlorpyrifos-methyl. The authors stated that the percutaneous route was considered to be the primary route of absorption based on the results of regression analyses, which showed a good correlation between actual hand exposure data (handwash) and urinary metabolite concentration. In conclusion, we state that the respiratory uptake of propoxur will not bias the conclusions regarding dermal absorption and the reduction effectiveness of protective clothing.

In our study, a good correlation was observed between handwash data and IPP excretion for harvesters without protective clothing ($R^2 = 0.87$), but this correlation decreased in the scenario with protective clothes, including cotton gloves ($R^2 = 0.25$). These observations indicate a relative increase in the contribution of percutaneous absorption of other exposed skin areas to IPP excretion. The same holds for applicators; however, it can be argued that a linear relationship exists between external dose and absorption (i.e., a fixed percentage of absorption that forms the underlying assumption of the linear regression analyses). Fick's first law of diffusion implies a maximum flux, so a further increase of the dermal area dose will not result in a higher uptake rate (Bos et al., 1996). The actual hand exposure of the applicators is largely determined by mixing and loading activities, which form only a small part of the total exposure period (approximately 5%).The lag-time necessary to reach a steady-state flux, although not known for propoxur, may exceed the short period exposure during mixing and loading.

In addition, the steady-state absorption flux is influenced by several conditions such as temperature and skin moisture. Meuling et al. (1991) showed an increase of percutaneous absorption of propoxur under conditions of occlusion. In a recently concluded volunteer study, the increase of IPP excretion was associated with an increase of relative humidity and the resulting increase of skin moisture contents (Meuling et al., 1997). An increase of 23 AU of skin moisture resulted in an increase of excretion from 13 to 63% of the dose applied at the forearm after 4 hr of exposure. The results of the regression analyses presented in Table 6 indicate an association between the difference in skin moisture due to protective clothing and the differences in IPP excretion between the clothing modalities. The observations of our study support the hypothesis that the change of skin variables under the conditions of protective clothing result in an enhancement of dermal uptake. Actual dermal exposure may be reduced by protective clothing, but the resulting dermal area dose will be absorbed more efficiently.

The skin moisture content of the forearms, "V" of the neck, and front of the head increased significantly under conditions of wearing protective clothing. In a pilot study, these skin areas reflected skin moisture changes of most other body areas rather well. Skin moisture contents of the hands showed large "within-" and "between-persons" variability. Apart from environmental factors, skin factors (e.g., temperature and moisture of the hands) are heavily affected by psychological factors (Havenith, 1996). Because the hands are important areas for actual skin exposure, results of skin moisture monitoring of the hands are considered to be relevant for estimates of dermal absorption but may be less relevant for assessment of the influence of protective gloves and clothing on thermal comfort.

Ojanen et al. (1994) evaluated the wearing comfort of different types of protective garments after field experiments, including the condition of the garments after laundering. Cotton suits were ranked as most comfortable. In our study, no field evaluation of the wearing comfort of the suits and gloves was conducted. Prior to the field study, different clothing configurations were evaluated for thermal comfort and ergonomics in a volunteer study (Havenith and Vrijkotte, 1993). A slight decrease was observed when light working clothing and no gloves were compared for both comfort sensation and ergonomics (e.g., skin moisture sensation). All subjects judged the clothing configuration acceptable. Finger and hand dexterity of nitrile gloves for the applicators and cotton gloves for the harvesters decreased 31% and 8%, and 23% and 0%, respectively, and were judged acceptable for good performance of the task.

In our study, no concurrent actual exposure of other body parts was assessed. In a previous field study, potential and actual exposure to methomyl by applicators using hand-held equipment was assessed using whole-body monitoring. The results showed that the median penetration of the active ingredient through cotton coveralls was only 2%, but 10% of the observations yielded a penetration of 10% or more (De Vreede et al., 1994). Based on comparison of exterior and interior patches, Davies et al. (1982) reported an average transmission of the pesticide through cotton coveralls of 0.7%, whereas the average transmission of pesticide through penetration for applicators wearing their own work clothing was 16.5%. Nigg and Stamper (1983) concluded that disposable coveralls would reduce actual whole-body exposure by 96%. Assuming an overall reduction of 95% by wearing coveralls and using the observed reduction rate for gloves, the contribution to the actual exposure of uncovered parts (i.e., the front of the head) would increase from 0.2 to 3.8%. Thus, the contribution from that particular area of the skin to dermal absorption might increase slightly.

Overall reduction efficacy (based on urinary excretion data) of a protective-clothing scenario observed in our study was relatively low compared to the observations of Chester et al. (1990b), who reported a reduction of approximately 80% for absorption of fluazifop-P-butyl. These results were based on comparison of the excretion of metabolites by applicators wearing gloves to a group of workers wearing no gloves. In our study, a similar approach would reveal a mean reduction of approximately 50% for applicators; however, median levels of approximately 40% in our study fit rather well within the results of the studies. In this case, overall reduction was also calculated from "within-worker" comparisons of internal exposure data. Davies et al. (1982) reported a reduction of internal exposure from protective clothing of approximately 50%, whereas Nigg and Stamper (1983) calculated a 24% overall reduction. The large range of overall reduction efficacy observed in our study was influenced partly by a variation in potential exposure for both scenarios. Working habits and individual factors related to dermal absorption contributed to the observed variability. Because the study design enabled "within-worker" comparisons of IPP excreted, individual kinetics did not contribute to the variation substantially. For risk evaluation purposes, the frequency distribution of the effectiveness of protective clothing based on estimates of internal exposure should be taken into account.

In conclusion it can be stated that the use of protective clothing will substantially reduce actual exposure of both applicators and harvesters. The observed decrease of the absorbed dose was not linearly related to the decrease of actual dermal exposure. The results suggest a more efficient absorption of the resulting actual exposure of propoxur while wearing protective clothing. The increase of skin moisture contents may be a key factor here and should be explored further. Because our results of overall exposure reduction based on biological monitoring fit into the ranges of other studies, it is indicated that the relative low reduction is not compound-specific.

Therefore, risk management utilizing protective clothing should be considered carefully, as estimation of the reduction of external exposure using protective clothing may overestimate the protective potential.

Acknowledgments

The research described in this paper was funded by the Dutch Ministry of Social Affairs and Employment. We want to thank the subjects for volunteering to participate in this study. For assistance in the field work we would like to thank Marieke Beelen, Bert Bierman, Marcel de Haan, Ceciel Lansink, and John Matulessy. Usha Soekoe, Roel Engel, and Lambert Leenheers are acknowledged for performing chemical analyses.

References

ACGIH (1985) *Particle Size Selective Sampling in the Workplace,* American Conference of Governmental Industrial Hygienists, Cincinnati, OH.

Aprea, C., Sciarra, G., Sartorelli, P., Desideri, E., Amati, R., and Sartorreli, E. (1994) Biological monitoring of exposure to organophosphorus insecticides by assay of urinary alkylphosphates: influence of protective measures during manual operations with treated plants, *Int. Arch. Occup. Environ. Health,* 66:333–338.

Bos, P.M.J., Brouwer, D.H., Stevenson, H., de Kort, W.L.A.M, and van Hemmen, J.J. (1996) Possibilities for the Assessment of Dermal Exposure Limits in the Occupational Environment, Report V96.464, TNO Nutrition and Food Research Institute, Zeist.

Brouwer, D.H. and van Hemmen, J.J. (1990) De effectiviteit van adembeschermingsmiddelen bij het werken met bestrijdingsmiddelen in de glastuinbouw. Beschrijving van de omstandigheden waaronder bestrijdingsmiddelen worden toegepast [*The Effectiveness of Personal Respiratory Equipment During Occupational Use of Pesticides in Greenhouse Crops. Description of the Conditions of Use of Pesticides*], Report 1990-23, Medical Biological Laboratory TNO, Rijswijk.

Brouwer, D.H. and van Hemmen, J.J. (1994) Fitting personal protective equipment (PPE) to the hazard: selection of PPE for various pesticide exposure scenarios in greenhouses, in *Book of Abstracts of the American Industrial Hygiene Conference & Exposition,* American Industrial Hygiene Association, Anaheim, CA.

Brouwer, R., Van Maarleveld, K., Ravensberg, L., Meuling, W., de Kort, W., and van Hemmen, J.J. (1993) Skin contamination, airborne concentrations, and urinary metabolite excretion of propoxur during harvesting of flowers in greenhouses, *Am. J. Ind. Med.,* 24:593–603.

Chester, G., Adam, A.V., Koch, A.I., Litchfield, M.H., and Tuinman, C.P. (1990a) Field evaluation of protective equipment for pesticide operators in a tropical climate, *Med. Lav.,* 81:480–488.

Chester, G., Loftus, N.J., Woollen, B.H., and Anema, B.P. (1990b) The effectiveness of protective clothing in reducing dermal exposure to, and absorption of, the herbicide fluazifop-P-butyl by mixer-loader-applicators using tractor sprayers, in *Book of Abstracts, Seventh International Congress of Pesticide Chemistry,* Vol. III, Freshe, H. and Kesseler-Smith, E., Eds., Conway, Hamburg.

Chester, G. (1995) Revised guidance document for the conducted of field studies of exposure to pesticides in use, in *Methods of Pesticide Exposure Assessment*, Curry, P.B., Iyengar, S., Maloney, P.A., and Maroni, M., Eds., Plenum Press, New York, pp. 197–216.

Cowell J.E., Danhaus, R.G., Kunstman, J.L., Hackett, A.G., Oppenhuizen, M.E., and Steimetz, J.R. (1987) Operator exposure from closed system loading and application of alachlor herbicide, *Arch. Environ. Contam. Toxicol.*, 16:327–332.

Crowse, J.L. , DeJonge, J.O., and Calogero, F. (1990) Pesticide barrier performance of selected nonwoven fabrics in laboratory capillary and pressure penetration testing, *Textile Research J.*, 60:137–142.

Davies, J.E., Freed, V.H., Enos, H.F., Duncan, R.C., Barquet, A., Morgarde, C., Peters, L.J., and Danauskas, J.X. (1982) Reduction of pesticide exposure with protective clothing for applicators and mixers, *J. Occup. Med.*, 24:464–468.

De Vreede, J.A.F., De Haan, M., Brouwer, D.H., van Hemmen, J.J., and de Kort, W.L.A.M. (1994) *Exposure to Pesticides. Part III. High-Volume Application to Chrysantemums in Greenhouses*, Report S131-4, Ministry of Social Affairs and Employment, The Hague, The Netherlands.

Dubelman, S., Lauer, R., Arras, D.D., and Adams, S.A. (1982) Operator exposure measurements during application of the herbicide diallate, *J. Agric. Food. Chem.*, 30:528–532.

Fenske, R.A., Hamburger, S.J., and Guyton, C.L. (1987) Occupational exposure to fosethyl-Al during spraying of ornamentals in greenhouses, *Arch. Environ. Contam. Toxicol.*, 16:615–621.

Fenske, R.A. (1988) Comparative assessment of protective clothing performance by measurement of dermal exposure during pesticide applications, *Appl. Ind. Hygiene*, 3:207–213.

Fenske, R.A., Blacker, A.M., Hamburger, S.J., and Simon, G.S. (1990) Worker exposure and protective clothing performance during manual seed treatment with lindane, *Arch. Environ. Contam. Toxicol.*, 19:190–196.

Freed, V.H., Davies, J.E., Peters, L.J., and Parveen, F. (1980) Minimizing occupational exposure to pesticides: repellency and penetrability of treated textiles to pesticide sprays, *Res. Rev.*, 75:159–167.

Havenith, G. (1996) Personal communication.

Havenith, G. and Vrijkotte, T.G.M. (1993) De effectiviteit van huidbeschermingsmiddelen bij het werken met bestrijdingsmiddelen in de glastuinbouw. Deel III. Comfort en ergonomie [*The Effectivity of Protective Clothing During Occupational Use of Pesticides in Greenhouse Crops. Part III. Comfort and Ergonomics*], Report IZF C-40, TNO Institute for Human Factors, Soesterberg.

Lander, F. and Hinke, K. (1992) Indoor application of anti-cholinesterase agents and the influence of personal protection on uptake, *Arch. Environ. Contam. Toxicol.*, 22:163–166.

Leenheers, L.H., Breugel, D.G., Ravensberg, J.C., Meuling, W.J.A., and Jongen, M.J.M. (1992) Determination of 2-isopropoxyphenol in urine using capillary gas chromatography and mass-selective detection, *J. Chromatography*, 578:189–194.

Leonas, K.K. (1991) Effect of pesticide formulation on transmission: a comparison of three formulations, *Bull. Environ. Contam. Toxicol.*, 46:697–704.

Machemer, L., Eben, A., and Kimmerle, G. (1982) Monitoring of propoxur exposure, *Stud. Environ. Sci.*, 18:255–262.

Maddy, K.T., Krieger, R.I., O'Connel, L., Bisbiglia, M., and Margetich, S. (1989) Use of biological monitoring data from pesticide users in making pesticide regulatory decisions in California. Study of captan exposure of strawberry picker, in *Biological Monitoring for Pesticide Exposure: Measurement, Estimation, and Risk Reduction*, Wang, R.G., Franklin, M., Honeycutt, R.C., and Reinert, J.C., Eds., ACS Symposium Series, 282, American Chemical Society, Washington, D.C., pp. 338–353.

Mark, D. and Vincent, J.H. (1986) A new personal sampler for airborne total dust in the workplace, *Ann. Occup. Hygiene*, 30:89–102.

Methner, M.M. and Fenske, R.A. (1994) Pesticide exposure during greenhouse applications. Part II. Chemical permeation through protective clothing in contact with treated foliage, *Appl. Occup. Environ. Hygiene*, 9:567–574.

Meuling, W.J.A., Bragt, P.C., Leenheers, L.H., de Kort, W.L.A.M. (1991) Dose-excretion study with the insecticide propoxur in volunteers, in *Prediction of Percutaneous Penetration, Methods, Measurements and Modelling*, 2nd ed., Scott, R.C., Guy, R.H., Hagraft, J., and Boddé, H.E., Eds., pp. 13–19.

Meuling, W.J.A., Franssen, A. Ch., Brouwer, D.H., and van Hemmen, J.J. (1997) The influence of skin moisture on the deraml absorption of propoxur in human volunteers: a consideration for biological monitoring practices, *Sc. Total Environ.*, 199:165–172.

Nigg, H.N. and Stamper J. (1983) Exposure of spray applicators and mixer-loaders to chlorobenzilate miticide in Florida citrus groves, *Arch. Environ. Contam. Toxicol.*, 12:477–482.

Nigg, H.N., Stamper, J.H., and Queen R.M. (1986) Dicofol exposure to Florida citrus applications: effects of protective clothing, *Arch. Environ. Contam. Toxicol.*, 15:121–134.

Nigg, H.N., Stamper, J.H., Easter, E.P., Mahonand, W.D., and DeJonge, J.O. (1990) Protection afforded citrus pesticide citrus pesticide applicators by coveralls, *Arch. Environ. Contam. Toxicol.*, 19:635–639.

Nigg, H.N., Stamper, J.H., Easter, E.P., and DeJonge, J.O. (1992) Field evaluation of coverall fabrics: heat stress and pesticide penetration, *Arch. Environ. Contam. Toxicol.*, 23:281–288.

Nigg, H.N., Stamper, J.H., Easter, E.P., and DeJonge, J.O. (1993) Protection afforded greenhouse pesticide applicators by coveralls: a field test, *Arch. Environ. Contam. Toxicol.*, 25: 529–533.

Oakland, B.G., Dodd, R.B., Schabacker, D.J., and Clegg, L.X. (1992a) Preliminary evaluation of nonwoven chemically treated barrier fabrics for field testing of protective clothing for agricultural workers exposed to pesticides, *Bull. Environ. Contam. Toxicol.*, 49:51–47.

Oakland, B.G., Schabacker, D.J., Dodd, R.B., and Ross, R.H. (1992b) The evaluation of protective clothing as chemical barriers for mixer/loaders and applicators in agricultural field tests designed to meet FIFRA GLP testing standards, in *Performance of Protective Clothing*, Vol. 4, McBriarty, J.P. and Henry, N.W., Eds., ASTM STP 1133, American Society for Testing and Materials, Philadelphia, PA, pp. 481–495.

Ojanen, K., Sarantila, R., Klen, T., and Kangas, J. (1992) Evaluation of the efficiency and comfort of protective clothing during herbicide spraying, *Appl. Occup. Environ. Hyg.*, 7:815–819.

Raheel, M. (1991) Pesticide transmission in fabrics: effect of particulate soil, *Bull. Environ. Contam. Toxicol.*, 46:845–851.

Scherle, W. (1970) A simple method for volumetry of organs in quantitative stereology, *Microscopy*, 20:57–60.

Snedecor, G.W. and Cochran W.G. (1982) *Statistical Methods*, 7th ed., The Iowa State University, Ames.

van Rooij, J.G.M., van Lieshout, E.M.A., Bodelier-Bade, M.M., and Jongeneelen, F.J. (1993) Effect of the reduction of skin contamination on the internal dose of creosote workers exposed to polycyclic aromatic hydrocarbons, *Scand. J. Work Environ. Health*, 19:200–207.

van Hemmen, J.J., Brouwer, D.H., Havenith, G., Kaaijk, J., and Brouwers, A.A.F. (1994) De effectiviteit van huidbeschermingsmiddelen bij het werken met bestrijdingsmiddelen [*The Effectivity of Personal Protective Clothing During Occupational Use of Pesticides in Greenhouse Crops*], TNO Centrum voor Arbeid, Leiden.

Wicker, G.W., Williams, W.A., and Guthrie, F.E. (1979) Exposure of field workers to organophosphorus insecticides: sweet corn and peaches, *Arch. Environ. Contam. Toxicol.*, 8:175–182.

chapter six

Operator exposure study with cyromazine 75 WP in water-soluble bags

J.R. Purdy

Contents

Introduction

Trigard® 75 WP is a new insecticide with a unique mode of action and a unique triazine structure. It is a solid formulated as a wettable powder and is packaged in water-soluble bags. The active ingredient in Trigard® has the common name cyromazine. Cyromazine is a triazine, but, unlike the well-known triazine herbicides, this compound has insecticidal properties and no herbicidal activity. Cyromazine has low mammalian toxicity and low vapor pressure. It is hydrophilic, so dermal penetration is expected to be minimal. This report focuses on the experimental design and sampling strategy for an operator exposure study with 15 volunteers who were experienced in application of crop-protection chemicals using field sprayers. The results of the study and conclusions regarding reduction of exposure are also presented. (See Figure 1.)

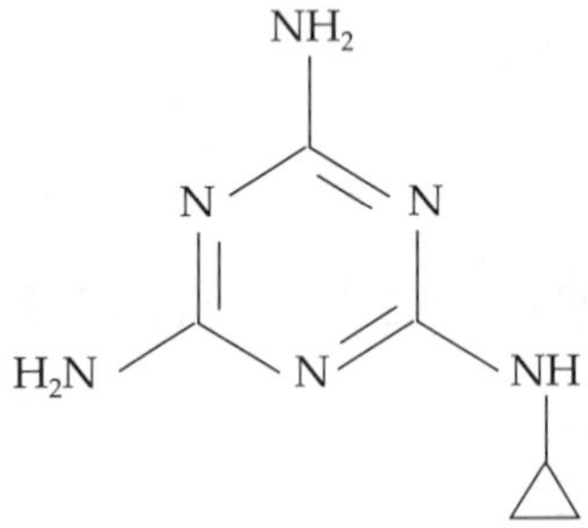

Figure 1 Cyromazine CGA 72662.

Study design

Exposure assessments are commonly done using computer modeling (e.g., Pesticide Handlers Exposure Database [PHED]). Computer modeling was not used for cyromazine 75 WG because of the limited information in the database on wettable powders packed in water-soluble bags at the time the study was planned.

A field operator exposure study was designed to accommodate the chemical properties and the use pattern involved in the treatment of potato crops for protection against the Colorado potato beetle. Fifteen farmworkers experienced in the use of tractor-drawn sprayers for application of crop protection materials to crops were selected. All were adult males. To avoid production of a large quantity of treated potato crop prior to registration of the product for sale, the trials were performed on stubble fields after harvest of winter wheat crops in a potato-growing area of southern Ontario. This allowed the use of typical farm spray equipment and a typical duration of exposure for a complete shift of work.

Because the objective of the study was to determine safety factors for workers under actual conditions of use, with all normal safety precautions, there was no deliberate attempt to apply a dose of product to the volunteers involved in the study. The sampling methods were designed to provide a complete measure of any exposure that occurred in the course of normal use of the product.

The study was also designed to meet Canadian Regulatory Guidelines in effect at the time the study was done. The most significant difference from U.S. Environmental Protection Agency's Subdivision U guidelines is in the definition of a replicate. The U.S. EPA guidelines consider a replicate to be a measurement of exposure of one operator during one work cycle; no activities are repeated. The Canadian Guidelines (as in Europe) consider a replicate to be a measurement of total exposure to an operator from all activities during a time period that is representative of a typical day of application work. This was taken to be approximately a half day, so only one set of samples was collected from each operator.

Using this criterion, each operator was asked to complete three cycles of loading the sprayer, mixing the test product, and applying the spray mixture, followed by cleaning the residual product out of the sprayer. For larger equipment, some operators filled the sprayer only partially, as this would allow for three cycles within the available land. The multiple cycles was considered important to ensure that the results include the effects of spreading the test material around the tractor and equipment and the effects of fatigue. Given the experience in this study, the three cycles resulted in 3 to 5 hr of exposure time, and each operator treated up to 40 hectares (100 acres) of land. This is typical of the amount of time spent spraying crops on a typical work day under Canadian agronomic conditions but also allows easy extrapolation to the extreme case of two work shifts on the same day or on two consecutive days, as a complete cycle of work activities was included.

The volunteer selection criteria were as follows:

- The volunteer had to be at least 18 years old.
- The volunteer had to be experienced in the use of crop protection chemicals.
- The volunteer had to be able to provide typical spray equipment.
- The volunteer had to understand the nature of the study and agree to participate.

The operators were asked to follow their normal work practices as much as possible. This ensured that the results represented the exposure likely to occur in the actual use of the product. All operators were cooperative and completed the required work.

Routes of exposure and sampling techniques

Due to the low volatility of cyromazine and the use of water-soluble bags for packaging the Trigard® formulation, the main routes of exposure were expected to be from direct contact with the product or spray mixture on contaminated surfaces. Previous experience with pesticides worker exposure studies indicated that exposure from vapors or spray mist would be a minor factor. This can easily be confirmed by the PHED or similar published sources; however, the extent of exposure from inhaling the product as dust is less well known. This route of exposure was also assumed to be minor, particularly with the use of water-soluble bag packaging. Given the low mammalian toxicity of cyromazine, the operators did not wear respiratory protection.

In designing the sampling, it was considered essential to focus on the dermal route of exposure, although air samples were also collected from the breathing zone of the operator. The air samples were collected at 2.0 L/min using polyurethane foam plugs (2.7 mm diameter × 2.5 cm deep). Two plugs were mounted in each sampling cassette and were analyzed separately. The absence of residues in the second plug was used to verify that no residues escaped capture in the first plug. This type of sampling medium captures both vapor and particulate matter. The cassette was connected to a battery-powered air-sampling pump mounted on a belt at the operator's waist. The air was collected continuously from the breathing zone of the operator at a calibrated flow rate throughout the exposure interval.

In designing the dermal exposure sampling, outer clothing typically worn by operators was considered essential, but to ensure uniform and reliable performance of the analytical method, the clothing was standardized by using the same clothing for all operators and providing the clothing at the start of the trial. The operators wore blue cotton cloth coveralls with long sleeves and legs, calf-length rubber boots, cotton work socks, and a baseball-type cap. No additional clothing was permitted outside the coveralls. Nitrile protective gloves were provided for use when handling the product or for other procedures where direct exposure to spray mixture might occur. The cotton coveralls were cut into four samples at the end of the exposure trial: arms, legs, front torso, and back torso. The socks were kept as a separate sample, but the boots were not sampled. A 25-cm diameter patch of cotton coveralls material with aluminum foil backing was used to cover the top of the cap and collect any residues which would have landed on the head portion of the hat.

During the trial, each operator was asked to wash his hands in distilled water with gloves on to obtain a glove wash sample after each load of spray mix was prepared and at the end of the trial, which represents proper use of nitrile gloves. At the end of the trial, after the washed gloves were removed, the operators washed their bare hands using Ivory soap and distilled water. The glove wash and bare hand wash were analyzed separately to assess the effectiveness of the gloves. A cotton swab moistened with soapy water was used to wipe residues from the operator's face and neck above the collar at the end of the trial. The swabs were also kept as a separate sample.

The outer layer of samples (gloves, coveralls, socks, face wash, hand wash, and hat) allowed measurement of the complete dose encountered on the outside of the protective clothing without any subsampling. This eliminated any uncertainty or error due to the highly variable deposition of residues across the body surface. This is the upper limit of the potential dose that could be encountered by the operator, and it is used to assess the effectiveness of the protective clothing and other preventive measures.

The operators were given cotton long underwear to wear under the coveralls and over regular underwear. The underwear sleeves from the elbow to the wrist were collected as one sample, and the remainder was taken as a second sample. The residues intercepted on the underwear plus the bare hand wash and face/neck wash were taken as a measure of the actual dermal exposure received by an operator wearing protective clothing. This measure, together with the air sample results, was taken as the total exposure experienced by the operator. Again, the entire sample was extracted for analysis, eliminating any subsampling error.

Analytical procedure

The cloth samples were collected immediately after the wash procedures were completed. They were packed in polyethylene bags and frozen in a portable freezer. The samples were kept frozen until they were analyzed. The liquid samples were packed in glass bottles and kept in insulated coolers with ice until analyzed. The cloth and cotton swab samples were extracted three times with deionized water. The extracts were acidified with HCl and loaded onto CUBCX cation exchange cartridges at pH 3. The cartridges were washed with 90% methanol water and eluted with ammonia/methanol. The concentrated extracts were filtered and analyzed by high-performance liquid chromatography (HPLC) with a Zorbax 300 SCX column and a diode array ultraviolet detector at 214.4 and 240.4 nm wavelength. The polyurethane foam plugs were extracted with acidified (pH 2) water and analyzed simi-larly. The washwater samples were acidified and loaded onto the ion exchange cartridges.

The primary sample types used for field spiking were freshly prepared soapy distilled water (soapy water), air filter cassettes set up with 2.0 L/min. of air flow, and foil-backed patches of underwear cloth with a cover flap of coveralls cloth. The spiking solution was applied to the underwear material and the coveralls patch was then folded down to cover the spiked area. The patch was then exposed to air and sunlight for the duration of the trial in an area upwind from the trial site. The washwater samples for spiking consisted of 50-mL samples of soapy water prepared by putting on latex examination gloves and washing with Ivory soap in deionized water prior to the trial in the same way the operator would wash his hands.

In the lab, the same primary sample types were used, and samples were also done to verify that the recovery from the coveralls alone was similar to that for the combined underwear plus coveralls patch and to verify that recovery from the individual types of cotton matrix was adequate. The results showed that recovery for the underwear was reduced by the large size of the garment compared to the patches used for the primary field and lab spiked samples. Recovery from the spiked cotton swabs was lower than expected but was still considered adequate (see Table 1).

Exposure results: total unprotected dermal exposure

The total unprotected dermal exposure (TUDX) is the sum of the exposure on coveralls, socks, hands, gloves, hat, face, and neck. This represents the maximum potential exposure of the operator, and this value and provides

Table 1 Recovery and Standard Deviation from Spiked Samples

Field-spiked and exposed for duration of trial (five replicates/trial)

Matrix	Percent recovery (S.D.)
Air filter (2.0 L/min air flow)	85 (4)
Underwear + coveralls cotton cloth patch	77 (6)
Soapy water	91 (10)

Lab-spiked samples

Matrix	Percent recovery (S.D.)	Number of samples
Air filter (2.0 L/min air flow)	91 (3)	8
Underwear + coveralls cotton cloth patch	93 (11)	9
Soapy water	94 (10)	11
Coveralls cloth alone	94 (10)	6
Hat patch	88 (7)	2
Facial swabs (cotton gauze/soapy water)	69 (10)	2
Socks (cotton)	99 (4)	2
Underwear (two sleeves, elbow to wrist)	97 (12)	2
Underwear (remainder of whole garment)	76 (2)	2

a base for evaluating the effectiveness of protective clothing as well as assessing the factors that contribute to exposure in the field. The ranges of exposure levels found in each sample matrix are listed in Table 2. The TUDX for each individual operator ranged from 159 to 17,078 µg/operator/day. The results show a wide range of values, but they followed the typical pattern seen with agricultural products applied as spray mixtures. The highest exposure areas were on the legs, the outside of the gloves, the arms, and the front of the coveralls. The hat patch and the face/neck wash had much lower values, and the socks were also relatively clean.

Four of the 15 operators had exceptional exposure compared to the other 11 which reveals improper use of protective gloves and/or handling contaminated surfaces with bare hands. The same four operators had high exposure on other areas and much lower safety factors which shows a clear link between poor work habits and higher exposure. The pattern of exposure confirms that direct contact with spray mixture on contaminated surfaces is the main route of exposure when using product in water-soluble bags.

Total actual dermal and respiratory exposure

The total actual dermal exposure (TDX) is the amount of material that would get onto the skin of the operator during a representative day of

Table 2 Total Unprotected Dermal Exposure Results
for 15 Operators Using Tractor-Drawn Sprayers

Sample matrix	Range of results (μg/operator/day)
Cotton coveralls	
Arms	10–2776
Legs	<11–11544
Front torso	<11–3490
Back torso	<10–290
Glove wash	117–5385
Hand wash	<8–2329
Face/neck wash	<0.5–6.7
Hat patch	<0.5–26.2
Socks	<2–8.8
Range of values of TUDX for individuals	159–17,078

Note: The values are not corrected for background or recovery. The TUDX values
were obtained by taking the sum of the results for each operator.

application work. In the sampling design, this includes the amounts on
exposed areas of bare skin (face/neck and bare hands), as well as the
amounts that penetrate protective clothing and are collected on the long
underwear. The original plan was to maximize analytical sensitivity by
taking the entire sample; however, it was observed that for some operators
the sleeves of the underwear were exposed at the wrists. The sleeves from
the elbow to the wrist were therefore analyzed separately, so that the
potentially higher residues in this area would not bias the results on the
rest of the body.

The TDX values were much lower than the TUDX values, indicating that
the protective clothing provides a substantial reduction in the amount of
residues reaching the skin of the operator. The same operators who had
higher overall exposure had higher actual exposure. The frequent occurrence
of nondetectable residues is notable and indicates that some operators were
able to keep exposure to levels at or near the limit of detection. The low
respiratory exposure (REX) confirms the validity of the assumption that
airborne exposure would be a minor route of exposure. (See Table 3.)

Safety factors

Two approaches were used for calculating the safety factors: the U.S. EPA
approach and the German model, which is one of the methods used in the
EEC.

$$\text{EPA safety factor} = \frac{\text{No effect level}}{\text{exposure (mg/kg body wt)}} = \frac{\text{NOAEL} \times 1000 \ \mu g/g \times 365/30}{\text{TDX/body wt (kg)}}$$

Table 3 Total Actual Dermal and Respiratory Exposure Results for 15 Operators Using Tractor-Drawn Sprayers

Sample matrix	Range of results (μg/operator/day)
Total actual dermal exposure (TDX)	
Cotton underwear	
Except forearm	4.5–72.6
Forearm	0.79–50.2
Hand wash	<8–2329
Face/neck wash	<0.5–6.7
Respiratory exposure (REX)	
Air-filter cassettes[a]	2.2–18.3
Range of values of TX[b] for individuals	15–2917

[a] Results are corrected for light work breathing rate of 8500 L per 8-hr day, or approximately 17.7 L/min).

[b] TX = TDX + REX.

Note: The values are not corrected for background or recovery. TX (total exposure) values represent the range of values of the total exposure obtained for each individual operator. The lower value represents one half the sum of the limits of detection. Individual operator results are provided in the next section.

The no observed adverse effect level (NOAEL) is obtained from the most relevant toxicity study. The lowest NOAEL observed on the toxicology studies was 1.5 mg/kg/day based on reduced weight gain in a 2-year chronic feeding study in rats. An estimate of no more than 30 days of use or handling per year was used in the calculation of the U.S. EPA safety factors.

$$\text{EEC safety factor} = \frac{1}{\text{TEQ}} = \text{Acceptable operator exposure level (AOEL) safety factor}$$

where TEQ = total exposure quotient:

$$\text{TEQ} = \frac{\text{TDX}}{\text{dermal AOEL}} + \frac{\text{Respiratory exposure}}{\text{inhalation AOEL} \times 2.0 \text{ L/min} \times \text{duration of trial (min)}}$$

The acceptable operator exposure level (AOEL) for each route of exposure is assigned from the no effect level (NOEL) in a specific toxicity test multiplied by a safety factor. The value for samples containing no detectable residues is assumed to be one half the limit of detection. For cyromazine, the seasonal use pattern indicates that the exposure is most comparable to the 21-day dermal exposure interval, and a value of 2000 mg/kg bw/day was taken as the dermal AOEL. The inhalation AOEL was obtained from a 28-day inhalation study with rats. As cyromazine is not a carcinogen, the safety margin used for calculation of the of the results using the EEC method

Table 4 Exposure Levels and Safety Factors
for Individual Operators (µg/day)

Trial no.	TUDX	TDX	REX	TX	EPA safety factor (000)	EEC safety factor
1	2492	38.3	5.4	43.7	474	217
2	981	44.2	2.7	46.4	259	116
3	7763	101.6	2.2	103.8	132	58
4	2249	13.0	2.2	15.2	1117	499
5	159	13.8	2.2	16.0	1035	473
6	4818	2899	18.3	2917	6	3
7	16,602	1129	5.1	1134	14	6
8	17,078	1502	4.0	1506	11	5
9	245	13.0	2.2	15.0	1269	532
10	321	20.3	2.2	22.5	582	264
11	1250	62.4	2.2	64.6	211	94
12	693	18.7	2.2	20.9	721	319
13	1785	17.5	5.1	22.6	568	270
14	8404	1327	2.2	1329	12	5
15	3130	99.3	2.7	101.5	143	63
Average					437	195
S.D.					433	190

Note: Values of 1/2 limit of detection were used to estimate the exposure in cases where no residues were detected. Trace values below the limit of detection were used (e.g., trials 4 and 10). The limit of detection for TDX was 30.5 µg, and the limit of detection for REX was 5.4 µg/day, making the limit of detection for REX 35.9 µg/day

was 25 for both the minor inhalation route and for the major dermal route. The calculations were done assuming 100% absorption of the dermal or inhalation dose, in the absence of actual measured values. Because actual dermal penetration is likely to be in the range of 5%, the safety factors are underestimated by a significant amount. The exposure and safety factor results are summarized in Table 4.

Discussion

The sampling systems were well accepted by the operators and easy to handle and provided good analytical recoveries. The polyurethane foam air-sampling cassettes allowed a high sampling flow rate with minimal flow resistance. The primary sections intercepted the residue efficiently, as no residues were found in the secondary sections above the limit of detection (0.5 µg). The cassettes were easy to handle in the spiking procedure, as well. Extension of the sampling to include multiple repetitions of the mixing/loading/application work cycle and inclusion of the sprayer cleaning operation at the end the trial provided valuable insights into the sources of exposure for workers involved in spray application.

Normally, the geometric mean exposure value is used, with an operator weight of 65 kg and a time period of a typical work day (5 hours) in the EEC calculation. When this is done, the safety factor for application of cyromazine on potatoes is 40. For comparison, the values by the EEC method for individual operators are presented in Table 4. All operators had large safety factors. The values were ≥6000 by the U.S. EPA method. In the EEC method, any AOEL value greater than 1 is considered acceptable, as the actual safety factor is built into the calculation. The U.S. EPA factors followed the same pattern of high and low values as the EEC results.

As expected, the inhalation exposure was a minor factor. Of 15 operators, 11 had no detectable respiratory exposure even though the limit of detection for the respiratory samples was lower than for the other sample types (0.5 µg/sample, or 2.2 µg per operator per day).

High TUDX values were correlated with high TDX values, which in turn represented a dominant influence on the safety factors. Four of the 15 operators had much higher exposure than the others, and four operators had low exposure, with either no detectable exposure within the protective clothing (trials 5 and 9) or only trace levels below the stated limit of detection (trials 5 and 10). The results for the remaining operators were clustered in the mid-range.

The dermal exposure was concentrated around the hands and showed that the main route of exposure was by direct contact with contaminated surfaces. Proper equipment and proper work practices were most important in eliminating this type of exposure. Operators who had equipment that required filling by a loose hose, manual folding and unfolding of the spray booms, or opening the filter drain manually under the sprayer for rinsing had high exposure on the arms, gloves, or hands. Improper use of the gloves led to secondary transfer of residues onto other surfaces, onto the legs, and onto the bare hands.

The inhalation exposure did not follow the same pattern as the dermal exposure, and the highest inhalation value may be due to direct transfer of residues onto the inhalation sample from the glove or hand of the operator. The water-soluble bags eliminate dust and contact with the concentrated product when the containers of product are being dispensed into the sprayer; however, the bags cannot prevent the type of exposure observed in this study, which was due to direct contact with the spray mixture. The proper use of protective clothing and the use of well-designed equipment, with power-assisted folding and unfolding of the spray booms, will greatly reduce the potential for exposure. For example, operator 2 was careful to use the nitrile gloves properly, and operator 5 had power-assisted spray booms.

The procedures most likely to cause high exposure included:

1. Filling the tank with a loose hose that might dip into the spray mixture as the tank fills.
2. Removing the main pump filter to drain and clean the spray tank without pre-rinsing the tank and lines with clean water.
3. Handling the nozzles or wet part of the spray boom with bare hands.

4. Removing the gloves in a manner that transfers residues onto clothing or bare hands.

Conclusions

1. The sampling design allowed for excellent acceptance by operators, maximized performance of the analytical method, and provided clear, conclusive results with minimal numbers of different sample types.
2. The strategy of whole-body sampling and elimination of any subsampling on clothing resulted in some large samples, which were more difficult to handle in the laboratory, but eliminated the large potential errors due to subsampling.
3. Safety factors were acceptable for all operators, even without considering the low dermal absorption of cyromazine.
4. Automated equipment reduces exposure by eliminating handling of contaminated surfaces.
   a. Power-assisted boom folding is commonly available.
   b. Automated rinsing of tank, lines, and filter is not commonly available but would greatly reduce exposure potential.
5. Dermal exposure by direct contact with contaminated surfaces was the main route of exposure; this could be greatly reduced by proper use and removal of nitrile gloves.
6. Exposure from dust was prevented by water-soluble bags, and exposure from spray mist/vapor was a minor route of exposure.
7. Operators with the lowest safety factors had high exposure due to specific, avoidable incidents, such as handling contaminated surface with bare hands or opening the main filter before rinsing out the spray mixture, thus getting large amounts of spray mixture on the gloves and transferring it to other parts of the body.

References

EEC Document No. 9553/93: Proposal for Council Directive Establishing Annex VI to Directive 91/414/EEC Concerning the Placement of Plant Protection Products on the Market.

Franklin, C. (1985) Occupational exposure to pesticides and its role in risk assessment procedures used in Canada, in *Dermal Exposure Related to Pesticide Use*, Honeycutt, R., Zweig, G., and Ragsdale, N.N., Eds., ACS Symposium Series No. 273, American Chemical Society, Washington, D.C.

Lundehn, J.-R. (1992) *Uniform Principles for Safeguarding the Health of Applicators of Plant Protection Products*, German Exposure Study Guidelines, Paul Parey, Berlin, pp. 76–79.

NOEL and NOAEL values are from unpublished reports within Novartis Crop Protection, Inc.

Pesticides Handlers Exposure Database (PHED) software, U.S. Environmental Protection Agency, Health and Welfare Canada, National Agricultural Chemicals Association (U.S.); software originally issued February 1992 with subsequent upgrades.

chapter seven

Pesticide exposure assessment: Jazzercize™ activities to determine extreme case indoor exposure potential and in-use biomonitoring

R.I. Krieger, T.M. Dinoff, and J.H. Ross

Contents

Introduction

In 1985, Berteau and Mengle (1985) of the California Department of Health Services and Maddy of the Department of Food and Agriculture conducted a preliminary review of pesticides used indoors. They noted several cases (six) from the Pesticide Illness Surveillance system in which illness was reported after structural pest control. Hypothetical exposure estimates for infants, children, and adults following label use for propoxur, DDVP, and chlorpyrifos were sometimes greater than toxic levels. In 1987, Berteau et al. (1989) reiterated the concern about the potential magnitude of indoor exposures, particularly for children.

Cotton-patch dosimetry (Durham and Wolfe, 1962) has been the principal method used to assess potential dermal exposure (PDE). PDE was calculated by first measuring the residue on each patch within a sector, assuming that residues were uniformly distributed. Uncertainty about the magnitude of these estimates of PDE results from uneven distribution of residues (Fenske, 1990), placement of patches on high-exposure regions of the body, and small sample area relative to the body's surface area of 19,400 cm^2 (Thongsinthusak et al., 1993). Biological markers of exposure, such as urinary metabolites, have been underutilized to evaluate estimates of absorbed daily dose (ADD) derived from PDE. When critical data are lacking, persons who must make decisions regarding use and exposure mitigation measures must utilize health-conservative default assumptions which inflate estimates of ADD. Limited efforts have been made to determine the relationship between ADD and data collected by external dosimeters (PDE).

Exposure assessments have become an essential element of contemporary risk assessment (NAS/NRC, 1983). The primary purpose of exposure assessment is to qualitatively and/or quantitatively determine exposure and absorbed dose associated with a particular use practice or human activity. Contemporary exposure assessors and risk managers place a high premium on accurate data obtained by monitoring chemical exposure scenarios and critical human activities or work tasks.

A 1982 guidance document of the World Health Organization (WHO, 1982) suggested use of loose-fitting, cotton, whole-body dosimeters (WBD) to overcome inefficient sample collection. The California Department of Food and Agriculture recognized the limited usefulness of patch dosimeters for determination of ADD in handler, harvester, and indoor pesticide exposure studies (Maddy et al., 1989). Whole-body dosimeters worn outside or inside standard work clothing may be a suitable means to quantitatively collect chemicals such as pesticides and some of their degradation products. The dose retained by the dosimeter combined with estimates of clothing penetration and percutaneous absorption may be used to estimate ADD.

Studies in indoor environments of dermal contact transfer required an estimate, and a tight-fitting whole-body dosimeter was adopted and initially considered as a surrogate for skin (Krieger et al., 2000). Contact with treated surfaces was limited to feet, hands, limbs, and torso. Standardized Jazzercize™ to represent daily human activities and maximum contact was incorporated into protocols for indoor exposure studies (Ross et al., 1990, 1991). Comparative studies will be reported elsewhere (Krieger et al., 2000).

This study was conducted to evaluate and compare ADD determined using whole-body dosimetry with results of two situational exposure studies conducted following use of a flea fogger under natural conditions. Chlorpyrifos was selected due to its general availability as a fogger for indoor flea control. Chlorpyrifos is poorly absorbed by the dermal route and readily cleared from the body in urine (Nolan et al., 1984). Trichloropyridinol was measured in 24-hr urine specimens of the volunteers and was converted to chlorpyrifos equivalents as a measure of absorbed dose. The study provided an opportunity to determine the relationship between intensive, high-contact dosimetry studies and the amounts of chlorpyrifos absorbed by two sets of adults who re-entered fogger-treated homes.

Method

Protocol

A protocol that described the work to be done and protected the rights of the volunteers in the dosimeter studies was developed for consideration by the Human Subjects Review Committee, University of California, Riverside. Paid volunteers were considered to be at "negligible risk." The approved protocol was designated HS-95-048. The document, including informed consent, was reviewed with participants at an orientation meeting prior to the study.

Exposure

The total-release foggers (6 oz) contained chlorpyrifos (1.000%), piperonyl butoxide (0.100%), pyrethrins (0.050%), N-octyl bicycloheptene dicarboximide (0.166%), and inert ingredients, including petroleum distillates (KRID™, K-Mart Corp.; Troy, MI). The aerosol cans were weighed before and after use to measure discharge.

Dosimeters

Volunteers wore either whole-body dosimeters (Sears' all-cotton union suits, cotton socks, and cotton gloves) or bathing suits, which exposed at least 75%

of the skin below the neck. Thirteen persons wore dosimeters, and 21 wore bathing suits (Krieger et al., 2000).

Venue

Two 180-ft^2 sections of a 4-year-old, floral print, nylon carpet of medium pile installed in meeting rooms of the University Club, University of California, Riverside, were treated with two foggers each, in accordance with label directions. The rooms were closed for 2 hr before being opened for 30 min prior to the scheduled 20-min exposure period. The same regimen was followed when foggers were used in a Sacramento apartment and a Riverside house trailer.

Chlorpyrifos availability

A short-handled modification of the CDFA (or California) roller was passed over an all-cotton dosimeter (1711 cm^2) (Ross et al., 1991) to collect transferable chlorpyrifos at the University Club. All results were expressed as μg chlorpyrifos per cm^2 of the collection surface.

Biomonitoring

Successive 24-hr urine specimens were provided by each volunteer. Collection in the dosimeter studies began 24 hr prior to the chlorpyrifos exposure (study day 0) and continued for 3 days based upon the 27-hr half-life of chlorpyrifos in humans (Nolan et al., 1984). Pre-exposure controls were obtained in all cases. Total urine volume was measured for each of the days, and 20- to 30-mL portions were stored frozen prior to analysis. The Sacramento collections were 48 hr and the Riverside collections were approximately 84 hr after re-entry.

Fogger home-use biomonitoring

Situational chemical exposure studies (SCES) (Krieger et al., 1991) are opportunities to index the magnitude of human exposures under natural conditions. The magnitude of absorbed chlorpyrifos was estimated in four adults, ages 21 to 25 (estimated). One male and female lived in a 784-ft^2 Sacramento apartment and used six foggers for flea control. The couple spent approximately 20 hr in the apartment during the weekend. Pre-use urine specimens and two successive 24-hr specimens (1000 to 1500 mL/day) were collected from each person. The second couple also used six foggers against fleas in their 1100-ft^2 Riverside house trailer. Complete urine specimens were collected for 3.5 days following use. The volunteers spent an estimated 14 to 18 hr within the trailer each day. Both sets of specimens were analyzed for 3,5,6-trichloropyridinol (TCP), and the first set was also analyzed for diethyl phosphate.

Analysis

The surface dosimeters (roller), union suit, socks, and gloves comprised the whole-body dosimeter. Each was analyzed separately for chlorpyrifos following ethyl acetate extraction. Volume was reduced, if necessary, by heating at 45°C in a fume hood and the samples were analyzed by gas-liquid chromatography using a thermionic detector. Acid hydrolysates of urine containing TCP were extracted with diethyl ether and derivatized with diazomethane. Samples were concentrated, dried with sodium sulfate, and cleaned up using a silica gel column. The product was eluted in hexane and analyzed by electron capture gas chromatography. Validation work showed a 95% recovery efficiency of TCP from urine. The limit of quantitation of TCP was 4 or 5 ppb.

Measurements

Potential dermal exposure (PDE) was the sum of the amount of chlorpyrifos retained by the dosimeter (socks, gloves, and union suit) during the 20-min exposure period. Absorbed daily dose (ADD) was the sum of chlorpyrifos equivalents measured in urine for days 2, 3, and 4. Home-use biomonitoring data are expressed as chlorpyrifos equivalents per day, as exposure continued throughout the test period.

Results

Transferable residues

Only a portion of the chlorpyrifos deposited upon the carpet can be transferred to the skin or clothing following normal human contact. An estimate of the dislodgeable fraction was obtained by rolling a 25-lb weight (modified CDFA roller) over a cotton dosimeter placed upon the treated carpet (Table 1). Pre-exercise residue levels were lower than post-exercise levels, and there was substantial variability. Potential dermal exposure was calculated assuming that the levels recovered on the cotton dosimeter following rolling would be equal to the levels expected on the skin following human activity. Complete (100%) absorption and entire body contact with treated carpet were assumed. The PDE was determined as follows:

$$(0.2 \ \mu g/cm^2 \times 20{,}000 \ cm^2)/70 \ kg = 57 \ \mu g/kg$$

Dermal and whole-body dosimeter transfer

The foundation for this work was laid in 1989 when we adopted a low-impact Jazzercize™ routine of about 20 min to estimate daily exposure following the use of a flea fogger. The exercises were chosen to provide maximal surface contact between the subject and the pesticide-treated carpet (Ross et al., 1990). In the present case, two groups of volunteers performed the exercises concurrently: one group wore cotton whole-body dosimeters

Table 1 CDFA Roller and Cotton Dosimeter Measurements
of Chlorpyrifos at Several Distances from a Fogger
Before and After Jazzercize™ Activity

| | | Dosimeter | |
| | Distance from fogger | Before[a] | After |
Fogger	(ft)	(μg/cm^2)	(μg/cm^2)
A	2	0.30	0.16
	4	0.06	0.04
	6	0.13	0.20
B	2	0.65	0.14
	4	0.43	0.045
	6	0.22	0.045
C	2	0.23	0.28
	4	0.053	0.032
	6	0.010	0.047
D	2	0.99	0.14
	4	0.064	0.034
	6	0.090	0.03
Mean		0.269	0.084
S.D.		0.297	0.079

[a] $(0.27 \pm 0.30 \ \mu g/cm^2)(0.096)(19,000 \ cm^2)/70 \ kg = 7 \pm 8 \ \mu g/kg$.

(WBDs) and the second group wore swimming suits to give maximum skin exposure. The cotton WBDs were chosen to normalize regional differences in surface contact and to increase the sample size relative to cotton-patch dosimeters (Durham and Wolfe, 1962).

In our studies, the WBDs accumulated an average of 13,757 μg chlorpyrifos (n = 13), which would represent an administered dose of about 200 μg/kg. If the dermal absorption rate of chlorpyrifos is 9.6%/24 hr in humans (Thongsinthusak, personal communication), then the ADD would be 200 μg/kg × 0.096 = 19 μg/kg. If clothing penetration is taken into account, the estimated (daily) ADD resulting from the intensive 20-min contact period was about 2 μg/kg.

The union suits retained an average of 8.8 ± 8.5 mg chlorpyrifos. Socks and gloves retained 1.6 ± 1.8 and 3.3 ± 3.8 mg, respectively. The distribution of dose (suit:socks:gloves) was very similar to that observed in the initial trial when the ratio of chlorpyrifos was 4:2:1 (Ross et al., 1990). The large standard deviations probably resulted from uneven distribution of chlorpyrifos on the treated carpets (Table 1).

Experimental biomonitoring

Both skin-exposed volunteers and those wearing WBDs were biomonitored for urinary clearance of trichloropyridinol. An unexpected finding was that before exposure (day 0) each 24-hr urine specimen contained measurable

Table 2 Summary of Human Chlorpyrifos Exposures Estimated
Using Whole-Body Dosimeters and Biomonitoring

Class	Chlorpyrifos equivalents (µg)			
Whole-body dosimeter (n = 13)				
External	13,758			
10% penetration	1376			
9.6% dermal absorption	132			

		Day		
	Control	1	2	3
Bathing suit (n = 21)	21	51	87	61
S.D.	18	52	81	51
WBD/biomonitored (n = 13)	15	12	25	17
S.D.	7	9	17	16

trichloropyridinol (equivalent to 4 to 82 µg chlorpyrifos). The cause of this finding was not determined, except in the case of the highest value (about 1 µg/kg), which probably resulted from previous home flea-control measures. Estimated exposures were calculated as the sum of urinary trichloropyridinol for days 2, 3, and 4. They were adjusted for background levels of TCP by averaging pre-exposure values and subtracting three times that value from the 3-day sum of urinary TCP.

The WBDs retained an average potential dermal exposure (PDE) of 13,757 µg chlorpyrifos. If clothing penetration is assumed to be 10% and dermal absorption 9.6% per 24 hr, then the absorbed dose would be 132 µg, and the absorbed dosage would be about 1.9 µg/kg. Biological monitoring of the 13 volunteers wearing cotton dosimeters indicated that the absorbed daily dose that penetrated the WBD and was absorbed was 2 µg chlorpyrifos equivalents/kg (Table 2).

The absorption of chlorpyrifos was also measured in persons who wore WBDs. The average amount of chlorpyrifos equivalents excreted during the 24-hr control period was 21 ± 18 µg. The subsequent daily average levels were 51 ± 52, 87 ± 81, and 61 ± 51 µg per person (n = 21). Adjusting each for average control, the adjusted ADD was 1.9 µg/kg. This finding strengthens use of the 10% clothing penetration factor and the 9.6% dermal absorption estimate for chlorpyrifos in humans.

Situational chemical exposure monitoring

Estimates of human exposure based upon a series of default assumptions and speculation yielded alarming results (Berteau et al., 1989) and a substantial amount of regulatory response by registrants. These situational exposure studies were intended to provide insight concerning the magnitude of human exposure following routine (or excessive) use of flea foggers. In the Sacramento

Table 3 Estimates of Human Exposure Derived from Whole-Body
Dosimetry, Biomonitoring, and Routine Use of Pesticide Fogger

Procedure	Dose (µg/person)	Absorbed dosage (µg/kg)
Whole-body dosimeter	13,758 (external)	1.9 ± 2.3[a]
Urine biomonitoring	132 (absorbed)	0.51 ± 0.55
Riverside house trailer (n = 2)	101, 147	1.3, 2.1
Sacramento apartment (n = 2)	8 (est.)	0.1

[a] (0.1 clothing penetration × 13.8 mg WBD × 0.096 dermal absorption)/70 kg bw = 1.9 µg/kg. Observed ADD estimate based upon WBD retention = (0.51 µg/kg)/(19 µg/kg) × 100 = 2.6%.

case, the estimated absorbed dose was initially assessed by monitoring diethyl phosphates, but none was detected (25 ppb) (Krieger, unpublished). Subsequently, TCP analyses were performed by DowElanco, and TCP was present at the detection limit. As a result, the daily exposure was estimated to be 8 µg equivalents, or an ADD of about 0.1 µg chlorpyrifos per kg.

The background TCP levels in urine of the volunteers were 47 and 63 µg chlorpyrifos equivalents per day prior to discharge of the foggers in their Riverside house trailer (Table 3). During the following three days, the ADDs (background corrected) were 1.3 µg chlorpyrifos equivalents per kg (80-kg male) and 2.1 µg chlorpyrifos equivalents per kg (70-kg female). The male was outside of the home 8 to 10 hr per day, and the female was outside about 5 to 7 hr per day. No correction was made, as their clothing was exposed to the insecticide treatment and doubtless contributed to their low level of exposure. Their ADDs were similar for each of the 3 days of monitoring, and these daily exposures are similar to those obtained with the 20-min, high-contact Jazzercize™ activity.

Discussion

When humans contact a chemical residue such as a pesticide on a treated surface, some of the deposit can be dislodged or transferred to skin or clothing. Ultimately, a portion of the amount transferred may be absorbed and constitute the absorbed daily dose (ADD). The ADD provides the most precise estimate of exposure that can be practically obtained for humans and has become the most useful expression of exposure for risk assessment and risk management.

Passive dosimetry, which proved useful for the pursuit of better workplace hygiene in agriculture during the past 40 years (Durham and Wolfe, 1962), yields unvalidated and excessive amounts of worker exposure (Krieger, 1996). Currently, our approach with respect to indoor and agricultural exposure assessments has been the evaluation of exposure estimates using well-known, studied chemicals to first understand the work task and at a later time develop chemical-specific studies as required in the regulatory arena.

In our study, volunteers wearing either WBDs or bathing suits performed a set of choreographed Jazzercise™ routines on chlorpyrifos-treated carpet. These exercises at a low work level were designed to represent a maximum daily surface contact and transfer of chemicals such as pesticides. Results were similar to exposures of two sets of adults who used insecticide foggers in their homes.

In conclusion, the whole-body dosimeter represents a relatively simple means to sample a transferable part of surface chemical residues. The dosimeter can be considered as a layer of clothing in close contact with skin, but an estimate of clothing penetration is required to estimate dermal exposure. As such, the WBD will intercept and retain 90% of the PDE and allow 10% to penetrate the skin. When dermal absorption is factored into the calculation, there is a remarkable similarity between absorbed dosage estimated from whole-body dosimetry and from urine biomonitoring of skin-exposed persons during programmed exercise and from biomonitoring of persons who had day-long exposure following use of foggers in their homes.

From the limited situational exposure monitoring done with persons who used chlorpyrifos foggers and subsequently re-entered their homes and had prolonged exposure to residues, the estimated ADDs from a Jazzercise™ protocol may represent maximum daily human contact and absorption. Inhalation and ingestion may be negligible contributors to ADD under normal conditions of fogger use, which results in surface deposits of insecticides. Those conditions are regarded as worst case, as homeowners typically use one fogger per room rather than estimates based on volume or floor space when seeking to control flea infestations.

The earlier estimates of Berteau et al. (1989) unquestionably overestimate the exposure potential of flea foggers. Intensive 20-min contact through the Jazzercise™ exercise (Ross et al., 1990) resulted in a low level of ADD measured by urinalysis or calculated from PDE (Table 3). Such close correspondence of the derived ADD from PDE and the measured ADD was not expected. Additionally, our situational monitoring demonstrates that under normal living conditions the daily (24-hr) exposure of humans following intensive use of flea foggers is similar to that which occurs during the Jazzercise™ period. Additional work is needed to validate these findings. If it is assumed that the transfer coefficient concept (dose = residue × time × transfer coefficient) can be applied (Krieger, 1996), in this circumstance a daily transfer coefficient of about 3000 cm²/hr can be calculated.

References

Berteau, P.E. and Mengle, D.M. (1985) *An Assessment of the Hazard from Pesticide Absorption from Surfaces*, Community Toxicology Unit, California Department of Health Services, Berkeley, May 17, 1985.

Berteau, P.E., Knaak, J.B., Mengle, D.C., and Schreider, J.B. (1989) Insecticide absorption from treated surfaces, in *Biological Monitoring for Pesticide Exposure*, Wang, R.G.M., Franklin, C., Honeycutt, R.C., and Reinert, J.C., Eds., ACS Symposium Series No. 382, American Chemical Society, Washington, D.C., pp. 315–326.

Durham, W.F. and Wolfe, H.R. (1962). Measurement of the exposure of workers to pesticides, *Bull. WHO*, 26:75–91.

Fenske, R.A. (1990) Nonuniform dermal deposition patterns during occupational exposure to pesticides, *Arch. Environ. Contam. Toxicol.*, 19:332–337.

Krieger, R.I. (1995) Pesticide exposure assessment, in *Proceedings of the Seventh International Congress on Toxicology*, Elsevier, Amsterdam, pp. 65–72.

Krieger, R.I., Bernard, C.E., Dinoff, T.M., Fell, L., Osimitz, T. G., Ross, J.I., and Thongsinthusak, T. (2000) Biomonitoring and whole body cotton dosimetry to estimate potential human dermal exposure to semivolatile chemicals, *J. Exposure Anal. Environ. Epidemiol.*, 10:50–57.

Maddy, K.T., Krieger, R.I., O'Connell, L., Bisbiglia, M., and Margetich, S. (1989) Use of biological monitoring data from pesticide users in making regulatory decisions in California: study of captan exposure of strawberry pickers, in *Biological Monitoring for Pesticide Exposure*, Wang, R.G.M., Franklin, C., Honeycutt, R.C., and Reinert, J.C., Eds., ACS Symposium Series No. 382, American Chemical Society, Washington, D.C., pp. 338–353.

National Research Council, National Academy of Sciences (1983) *Risk Assessment in the Federal Government: Managing the Process*, National Academy Press, Washington, D.C., pp. 1–191.

Nolan, R.J., Rick, D.L., Freshour, N.L., and Saunders, J.H. (1984) Chlorpyrifos: pharmacokinetics in human volunteers, *Toxicol. Appl. Pharmacol.*, 73:8–15.

Ross, J., Thongsinthusak, T., Fong, H.R., Margetich, S., and Krieger, R. (1990) Measuring potential dermal transfer of surface pesticide residue generated from indoor fogger use: an interim report, *Chemosphere*, 20:349–360.

Ross, J., Fong, H.R., Thongsinthusak, T., Margetich, S., and Krieger, R. (1991) Measuring potential demand transfer of surface pesticide residue generated from indoor fogger use: interim report II, *Chemosphere*, 22:975–984.

chapter eight

Uniform principles for safeguarding the health of workers re-entering crop growing areas after application of plant-protection products

B. Krebs, W. Maasfeld, J. Schrader, R. Wolf, E. Hoernicke, H.-G. Nolting, G.F. Backhaus, and D. Westphal

Contents

Abstract

Any possible risk to users while spraying plant-protection products must be evaluated before its authorization. When it is necessary for workers to re-enter crops shortly after their treatment, they may be exposed to the plant-protection product through contact with the spray deposit; therefore, this possible means of exposure must be evaluated when a plant-protection product is to be used. This chapter discusses the procedure for and some examples of the assessment of re-entry exposure and provides instructions for placing protective measures on product labels if dermal and inhalative exposure is relevant.

Introduction

Plant-protection products are used in agriculture, horticulture, and elsewhere in order to prevent major yield and quality losses. They are an integral part of various types of crop maintenance measures. This means that other maintenance activities may make it necessary to re-enter treated areas relatively shortly after application (i.e., within the normal time frame for major plant-protection activities). The type of work to be done and the point of time for re-entering relative to the time of application of a plant-protection product may vary from crop to crop.

In general, plant-protection products are biocidal active substances and are therefore by nature toxic to target organisms. At least some of them are also toxic to humans; therefore, the safe use of plant-protection products presupposes, among other things, an evaluation of worker exposure during re-entry, an adequate risk assessment on the basis of the various practical scenarios in agriculture and horticulture, and, if necessary, specific instructions for worker protection on the product label.

Those activities are generally conducted during the course of official registration of a plant-protection product. Because, according to Council Directive 91/414/EEC, the registration procedures and evaluation criteria are currently harmonized between the member states of the European Union, the objective of this paper is to contribute to the discussion about a future Union-wide evaluation and assessment scheme for re-entry exposure.

Determination of re-entry exposure

Re-entry exposure will be predominantly via the dermal route. Exposure data indicate that, in general, inhalation exposure is only important during a relatively short period after application (e.g., in field crops only during the time when the spray is drying or in greenhouses within a few hours after application). Therefore, specific label instructions for adequate protection from health hazards via inhalation are generally confined to short-term restrictions on re-entering a treated crop or greenhouse.

In general, residues on the crop foliage dissipate relatively slowly, depending on the physical and chemical properties of the applied active substance. As most maintenance activities include frequent contacts with the foliage of the crop, dermal exposure is considered to be by far the most important exposure route. The common methodologies for determination of foliar residues and dermal exposure have been described elsewhere (Gunther et al., 1977; Iwata et al., 1977; Popendorf and Leffingwell, 1982; van Hemmen, 1993). This paper will concentrate on compilation and evaluation of the residue and exposure data for regulatory purposes.

Compilation of residue data for various exposure scenarios

Research has shown a clear relation between the amount of dislodgeable residues on the crop and the level of dermal exposure (van Hemmen, 1995). The maximum amount of dislodgeable residues is found immediately after application and depends on:

- Application rate
- Extent of residues remaining on foliage from previous applications, especially in ornamentals
- Crop habitat — total size of the foliage compared to the ground surface area, as indicated by the LAI, defined as the total foliage (one-sided) surface area of a crop divided by the ground surface area on which the crop grows (van Hemmen, 1993)

In the case of only one application and after normalization of the residue concentration to a unit rate (e.g., 1 kg active substance per hectare), multiple experiments in a single crop should result in a fairly similar dislodgeable foliar residue (DFR) level. If the LAIs are comparable within a certain crop group or even between different crop groups, then the respective DFR data may be compiled and evaluated as a single set of data. Furthermore, as the transfer of residues from the crop to the clothes or skin of the worker is more or less independent of the kind of product applied, the level of worker exposure will depend only on the intensity of contact with the foliage. This, again, is determined by the nature and duration of the maintenance activity to be carried out during re-entry.

It is therefore advisable to group the various crop habitats and maintenance activities into "re-entry scenarios" and to determine whether standard values for the initial DFR shortly after the first application and generic transfer factors for the level of dermal exposure for each scenario can be developed. Investigations to this end have been carried out over the last two decades, primarily in the U.S. The generic transfer factors for a number of scenarios, in particular, have been developed and discussed in detail (Krieger et al., 1992; Popendorf and Leffingwell, 1982).

Apart from crops (mostly high growing) where re-entry is or may be of concern, there are a number of other crops for which no regular re-entry into treated areas needs to be considered. This applies especially to highly mechanized field crops such as cereals, maize, or sugarbeets; therefore, it seems advisable, for adequate worker protection purposes, to focus on re-entry scenarios that are seen to be of greatest relevance and to see whether general instructions on the product labels might be sufficient in other cases where re-entry is only occasional. The re-entry scenarios considered to be relevant specifically for European agricultural and horticultural conditions are listed in Table 1. Besides the relevant crops and crop groups and the growing conditions (indoor/outdoor), the table also contains the types of important maintenance activities (work tasks) for the various crops or crop groups.

The number of scenarios may be even further reduced if, in a first approach, "worst-case" considerations with respect to the expected exposure level are taken into account:

1. Thinning activities are considered to result in less or, at most, comparably intensive contacts with foliage than pruning. In addition, more or less the same body parts of the worker will be exposed to the foliage during both types of activities; therefore, exposure calculations for pruning will also cover the worst-case exposure assumptions for thinning.
2. Pruning of pome fruit crops will lead to DFR levels comparable to those for stone fruit crops; therefore, a common database for both crop groups (for tree fruits) should be compiled. Further generic sets may be developed for fruit vegetables, grapes, and ornamentals.
3. Harvesting activities may contain common elements of handling practices which are similar for the various crops. If so, the respective data may be compiled to a single data set. Bearing this in mind, separate data sets should be compiled in a first approach only for the following groups of crops:
   a. Citrus, pome fruits, and stone fruits with large fruits (e.g., peaches)
   b. Fruit vegetables, bushberries, and stone fruits with small fruits
   c. Grapes
   d. Strawberries
   e. Head cabbage, head lettuce, and ornamentals, including pot flowers
   f. Tobacco
4. According to Table 1, dermal exposure during post-harvest activities results primarily from sorting and packaging (bundling). Especially in the case of sorting, workers must handle the harvested commodity with their hands, most often in a way similar to during harvesting but without any contact with foliage. Therefore, in a first attempt, these post-harvest exposures should be considered, at the most, as

Table 1 Examples of Re-entry Scenarios in Europe

			Type of post-application activity (work task)			
No.	Crop/crop group[a]	Green-house/ field[b]	Thinning	Pruning	Harvesting	Post-harvest activities
1	Citrus (oranges)	F	—	—	X	Sorting, packaging
2	Pome fruit (apples)	F	X	—	X	Sorting, packaging
3	Stone fruit	F	X	X	X	Sorting, packaging
4	Grapes	F	—	X	X	Packaging
5	Strawberries	F	—	—	X	—
6	Strawberries	G	—	—	X	—
6a	Fruit vegetables (tomatoes, cucumbers, etc.)	G	—	X	X	Sorting, packaging
7	Bushberries (currants)	F	—	—	X	—
8	Fruit vegetables (tomatoes, cucumbers, etc.)	F	—	X	X	Sorting, packaging
9	Head cabbage/ head lettuce	G/F	—	—	X	—
10	Ornamen-tals/pot flowers	G/F	—	X	X	Sorting, packaging
11	Tobacco	F	—	—	X	—

[a] Recommended test species for re-entry investigations.

[b] F = field; G = greenhouse.

comparable to exposure during harvesting. Accordingly, the specific instructions for worker protection during harvesting should be extended to post-harvest activities. An exemption toward less-stringent requirements for personal protection might be the mechanical treatment of fruits in a cleaning bath.

If the outlined procedure is followed, the data on the transfer of residues from foliage to the clothes or skin of workers will be limited to data sets for four different pruning and six different harvesting scenarios. In this framework, the available data in literature should be evaluated and the transfer factors for the various scenarios compiled. The range of transfer values will show whether the above outlined grouping of scenarios is appropriate or whether a revision is advisable.

Tiered approach to evaluation of re-entry exposure

As outlined above, the risk of workers during re-entry of a treated crop depends on:

1. Need to re-enter a treated area
2. Toxicity of the compound used
3. Level of exposure

It is advisable, then, in a tiered approach to concentrate first on crops and activities (scenarios) that are considered to be relevant with respect to the expected level of exposure and to exclude those not relevant. Second, whether or not the toxicological properties of the product may lead to general restrictions on re-entry should be investigated. If both the likelihood of re-entry and the hazard due to the toxicity of the compound cannot generally be neglected, a risk assessment over several steps should be carried out. The assessment may be based on surrogate data and "worst-case" assumptions at first and then refined, if necessary. One possible approach to a tiered evaluation procedure is presented in Figure 1.

Tier I

The procedure starts out assuming there is a likelihood of regular re-entry into a certain crop where the product of concern is — or will be — registered for activities resulting in a fairly high level of exposure. For this purpose, the crops and crop groups and the growing conditions (outdoor/indoor) listed in Table 1 may provide sufficient guidance. Exposure situations for other crops and crop groups will either be covered by the measures taken under Tier II or by application of the minimum label instruction, as discussed below. As a matter of general working hygiene, it is generally recommended to put the minimum label instruction on all labels of those plant-protection products for which application leads to wetting of the plant surface.

Tier II

This step contains an initial evaluation of possible risk to workers due to the toxicity of the plant-protection product. The evaluation is based on the classification and labeling requirements established within the European Community. In order to address the specific need to safeguard the health of workers during re-entry to treated crops, the general safety advice (S-phrases) has to be specified in detail. Delineation of specific instructions for workers regarding classification and labeling is provided below. If no classification or

Tiered approach to evaluation of re-entry exposure and
assignment of specific instructions for worker exposure

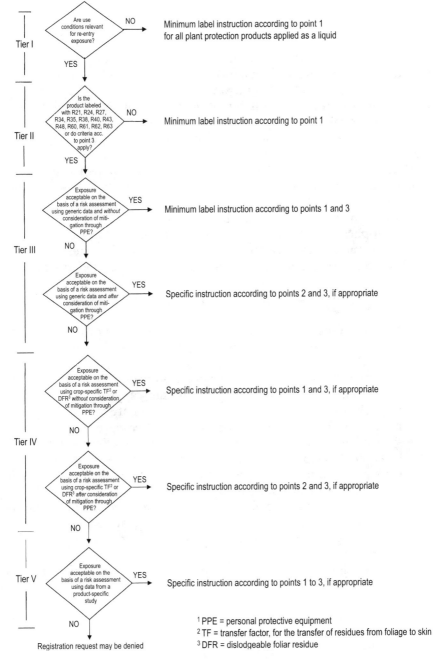

Figure 1 Tiered approach to evaluation of re-entry exposure and assignment of specific instructions for worker exposure.

labeling concerning health protection is required, only the minimum label instructions need to be applied; otherwise, a risk assessment according to Tier III is required.

Tier III

The quantitative estimation of available residue data from the published literature may be investigated in a further step, and these data may then be used to calculate the initial residue on the day of application (*after* drying of the spray mixture). An assessment of literature data by Schrader (1994) has revealed an initial dislodgeable foliar residue (DFR) below or around 1 µg/cm^2/kg active substance (a.s.)/ha, if the residue values are corrected for a standard rate of 1 kg a.s./ha.

The assumption of a DFR of 1 µg/cm^2/kg a.s./ha applied as the upper residue level on the day of application is supported by the following facts:

1. The theoretical initial concentration on leaves is 10 µg/cm^2, based on an application rate of 1 kg/ha, no loss of spray to the ground or into the air, and a leaf area index (LAI) of only 1, which means that the total foliar surface in a treated area is equal to the ground surface.

$$1 \text{ kg/ha} = 1 \times 10^9 \text{ µg}/1 \times 10^8 \text{ cm}^2 = 10 \text{ µg/cm}^2$$

2. If it is assumed, more realistically, that both sides of the leaves are sprayed, the residue concentration will be only in the range of 5 µg/cm^2.
3. Moreover, the LAI in nearly all crops is in reality above 2, primarily in the range of 3 to 5, especially in high growing crops. Correspondingly, the foliar residue will be lower by a factor of 3 to 5 (i.e., 5 µg/cm^2 divided by 3 to 5 = 1.66 to 1 µg/cm^2). If the crop does not cover the sprayed area completely (e.g., in row crops), then the theoretical calculation of the foliar residue still holds true, at least for conventional spraying equipment, as a proportional part of the spray mixture will not reach the foliage but will fall down to the soil.
4. Depending on the crop and the physical and chemical properties of the active substance, various portions of the applied amount will evaporate during the drying of the spray mixture and within the first few hours afterwards.
5. Finally, it must be taken into account that only a part of the total foliar residue will be dislodgeable.

For some time in the past, a single transfer factor of 5000 for the transfer of residues from foliage to the clothes or skin of the worker was used (Popendorf and Leffingwell, 1982). After generation of further data, it seems to be necessary to consider various transfer factors for the various work tasks, as outlined above. But, in a first attempt to calculate the dermal exposure, a default value in the range of the upper transfer factor for a work task with the most intensive contact to the foliage may be used, according to literature a value of 30,000. In case of less-intensive foliage contact (e.g., thinning activities), other transfer factors down to 5000 might be more appropriate. If this calculation does not reveal a potential health risk, then no further actions will be necessary. The calculation procedure on the basis of default values may than be as follows:

Dislodgeable foliar residue (DFR):	1 (μg/cm^2/kg a.s./ha)
Worst-case transfer factor (TF), double sided:	30,000 (cm^2/person $\times$ hr)
Workrate per day (A):	8 (hr/day; to be considered as "worst case," as it is only relevant if the residue is stable over 8 hr)
Penetration through clothes and gloves (P):	5% (0.05)
Application rate (R):	Variable (kg a.s./ha)

Example calculation
Dermal exposure (D):

$$DFR \times TF \times A \times (P) \times R$$

D:

$$1 \times 30.000 \times 8 \times (0.05) \times R$$

D (R = 1kg/ha):

$$240(12)\frac{mg}{person \times day}$$

The exposure value thus yielded provides a measure of the skin exposure with and without consideration of a protective garment and gloves (personal protective equipment = PPE), and may be taken directly for comparison with appropriate data from relevant toxicity studies for assessment of the risk via the dermal route.

Tier IV

If the risk assessment based on generic data and default values (after consideration of protective measures) does not provide a sufficient margin of safety, compound-specific data on the DFR will have to be developed. These data will afterwards be used in the above outlined procedure instead of generic data from literature.

Tier V

If a further refinement of the determination of exposure is necessary, an exposure study using the product of concern and conducted under conditions of real practice might be required. The exposure study may include passive dosimetry or biomonitoring, depending on the properties of the active substance and the data on metabolism and toxicokinetics in mammals.

Definition of measures for reducing exposure: label instructions

(See Figure 1.) If re-entry into certain crops, regularly or occasionally, cannot be excluded, but a "worst-case" risk assessment shows that exposure remains below the tolerable level even without specific skin protection measures, it will be sufficient to include the minimum label instruction, point 1:

> *Do not re-enter treated areas/crops before the spray deposit is completely dry.*

The inclusion of the minimum label instruction might also be considered as a matter of general working hygiene for all products for which application leads to wetting of the plant surface.

If the classification and labeling of the product according to Tier II indicate that body protection is necessary, the following specific instructions, points 2 and 3, should be included on the label, unless the risk assessment according to Tiers III through V indicates otherwise. Point 2:

> *Do not re-enter treated areas/crops on the day of application unless wearing the personal protective equipment specified for the application of the product. Re-entry activities on or in treated areas or crops may not be carried out during the first 24 hours after application of the product. Wear a protective garment and protective gloves during the first 48 hours after application.*

Note that if the risk assessment indicates the need for specific body protection for more than 48 hours after application, then the specific instructions above should be duly modified.

In the case of plant-protection products that are used in greenhouses/closed rooms:

1. With the risk phrase R42 ("may cause sensitization by inhalation"),
2. With an inhalation toxicity (LC_{50}, 4 hr) < 20 mg/L air,
3. Applied in the form of vapor, mist, or fumes, or
4. Requiring the use of respiratory protection during application,

the following specific instruction, point 3, shall be included additionally on the label:

> *Ensure thorough ventilation of greenhouses/closed rooms before re-entry.*

Note that, as a matter of precaution, greenhouses/closed rooms should be thoroughly ventilated after each treatment, regardless of whether the product is classified and labeled in regard to health protection or whether the risk assessment reveals a specific exposure risk for workers. In a similar sense, and whenever possible, crop maintenance activities should be finished before plant-protection products are applied.

Concluding remarks

Because the whole idea of a tiered approach of the kind outlined above is in its initial stages, it will have to be validated and discussed further and will in all probability need to be refined afterwards. The aim here is to introduce the idea of a stepwise approach to the assessment of the risk to re-entry workers. The outlined procedure should be used to calculate the dermal re-entry exposure for real examples of rather dermally toxic compounds in order to gain experience with the recommended procedure.

References

Gunther, F.A., Iwata, Y., Carman, G.E., and Smith, C.A. (1977) The citrus re-entry problem: research on its causes and effects and approaches to its minimalization, *Residue Rev., 67*:1–139.

Hoernicke, E., Nolting, H.-G., and Westphal, D., Fachausschuss Anwenderschutz (IVA) (1998) Hinweise in der Gebrauchsanleitung zum Schutz von Personen bei Nachfolgearbeiten in mit Pflanzenschutzmitteln behandelten Kulturen (worker re-entry), *Nachrichtenb. l. Deut. Pflanzenschutzd.,* 50.

Iwata, Y., Knaak, J.B., Spear, R.C., and Foster, R. J. (1977) Worker re-entry into pesticide-treated crops. I. Procedure for the determination of dislodgeable pesticide residues on foliage, *Bull. Environ. Contam. Toxicol.,* 18:649–655.

Krieger, R.I., Ross, J.H., and Thongsinthusak, T. (1992) Assessing human exposure to pesticides, *Rev. Environ. Contam. Toxicol.,* 129:1–15.

Popendorf, W.J. and Leffingwell, T. (1982) Regulating organophosphate residues for worker protection, *Residue Rev.,* 82:125–201.

Schrader, J. (1994) Personal communication.

van Hemmen, J.J. (1993) Re-entry Exposure and Product Development Pesticides and Greenhouse Crops: An Example, paper presented at one-day seminar, Agrochemical Occupational Risk Assessment, Brussels, June 15, 1993.

van Hemmen, J.J. (1997) Re-entry Exposure to Pesticides in European Cultures (unpublished).

chapter nine

Modeling re-entry exposure estimates: techniques and application rates

D.H. Brouwer, M. de Haan, and J.J. van Hemmen

Contents

Abstract

A study was conducted to validate some basic assumptions made in predictive models for estimating worker exposure to pesticides during re-entry. Leaf samples were collected prior to and following application of the pesticides abamectin, thiophanate-methyl, and methiocarb (in seven commercial greenhouses for the cultivation of carnations) using either a high-volume (HV) or low-volume (LV) method. During each consecutive re-entry, both respiratory and dermal exposure to the pesticides were assessed for a period of 4 weeks, starting from the moment of HV application. During re-entry, the following determinants of exposure were also evaluated: actual dislodgeable foliar residue (DFR) in the two major contact zones, number of flowers harvested, moment of re-entry (days after application), and crop volume index. The study investigated the relationship of the transfer factor (TF) — the proportion of DFR of a pesticide available for transfer from the treated foliage to the worker by contact — to re-entry time (days after application). Relationships between exposures and determinants of exposure were analyzed using multiple linear regression.

A linear relationship between the average increase of DFR (ΔDFR) and application rate for both application techniques was observed; therefore, the basic assumption made for exposure modeling seems acceptable. Calculated transfer factors for harvesting carnations using all dermal exposure data were 2300 cm^2/hr. A transfer factor of approximately 1600 cm^2/hr for methiocarb was observed. No significant relationship between dermal exposure and DFR was observed for thiophanate-methyl. For three out of 11 workers, a significant relationship between TF and re-entry time was found, whereas for the other workers a decreasing trend in transfer factors was observed.

Half-lives of two pesticides were estimated using the assumption that the decay was a first-order process. Half-lives were 28 ± 8.5 days and 12 ± 3.4 days for thiophanate-methyl and methiocarb, respectively. Crop volume was not a significant determinant of exposure using a grid count method. This was probably due to a large variation in the results with this method.

In conclusion, the basic assumptions of the re-entry model (i.e., a linear relationship between application rate and ΔDFR and a first-order decay of the DFR) were confirmed. The relationship between the transfer factor and level of DFR and re-entry time should be explored further. Including foliage surface area or crop density may be a refinement of the model; however, the estimation of crop volume should be improved before its influence on the exposure processes can be ascertained.

Introduction

Pesticides are widely used in agriculture. Occupational exposure to a pesticide should be evaluated before authorization of the pesticide for agricultural application; therefore, it is necessary to consider, among other factors, the possible health risks for the applicators and, in the case of re-entry, other workers. In the Dutch authorization procedure, when no actual exposure data are available, exposure is estimated using the following descriptive models: the Dutch model, based on a critical review of exposure levels found in open literature (van Hemmen, 1992, 1993; van Hemmen et al., 1995), the U.K. model (MAFF, 1986, 1992), and the German model (Lundeh et al.,1992). These models use generic exposure data which are adjusted for the actual pesticide and its label prescriptions (e.g., application rate). All models are thus based on the basic assumption that there is a linear relationship between application rate and level of exposure.

Re-entry exposure can be considered to be the result of a chain of processes ending with the transfer of pesticide residues to the worker. Popendorf (1985) assumed a linear relationship between dermal exposure (DE) and the amount of residue available to transfer from the foliage to the worker. This is called the dislodgeable foliar residue (DFR):

$$DE = k \times DFR \tag{1}$$

The factor k (the transfer factor) was defined as a crop- and task-specific factor and is defined as the slope of the line that fits dermal exposure levels (g/hr) and corresponding levels of DFR (g/m²) on the crop (i.e., the regression coefficient of DFR). The DFR, according to the procedures described by Iwata et al. (1977), was considered to be a good estimate of source strength for re-entry exposure.

The actual amount of the DFR at the time of re-entry may differ from the initial amount of DFR, depending upon the decay rate of the pesticide and the elapsed time since application. The process of decay may in many cases be considered as a first-order process (Willis and McDowell, 1987). These processes are given by the equations:

$$DE_i = DFR_{i,t} \times TF_m \times T_m \tag{2}$$

where:

DE	=	dermal exposure (g/day)
DFR	=	dislodgeable foliar residue (g/m²)
i	=	i-th pesticide

t   =   t-th day after application
TF   =   transfer factor (m^2/hr)
m   =   m-th task
T   =   duration of re-entry (hr/day)

and

$$\ln DFR_{i,t} = \ln DFR_0 + k_i t \tag{3}$$

where:

DFR$_0$ =   initial DFR (day 0 after application)
k   =   first-order reaction constant

and

$$k = -\ln 2(t_{1/2})^{-1} \tag{4}$$

where:

$t_{1/2}$ =   half-life (days)

It is assumed that the initial amount of DFR is linearly related to the applied amount or application rate (g active ingredient [a.i.] per ha), the surface area of the foliage (m^2), and the efficiency of the interception of the applied pesticide (Bates, 1990; Brouwer et al., 1994). Brouwer et al. (1994) stated that the level of interception depends on the application technique and the crop density (leaf area index, LAI). In general form:

$$\Delta DFR = \{AR_i \times (V \times A)_a\} \times 1/LAI_j \times I_{j,a} \tag{5}$$

where:

ΔDFR =   increase of DFR
AR   =   application rate (g a.i./100 L)
i   =   i-th pesticide
V   =   volume rate (L/m^2)
A   =   ground surface area (m^2)
a   =   a-th technique
LAI   =   leaf area index (m^2/m^2)
j   =   j-th crop
I   =   percentage of interception

Dermal exposure during repeated re-entry is assumed to differ from the first re-entry only because of dissipation of the pesticide (Popendorf, 1985). However, there are indications from previous studies in the cultivation of chrysanthemums that, when DFR is low (when time between the last application and re-entry is long), the transfer factors are different from transfer factors based on situations in which re-entry times are rather short (Veerman et al., 1994). This may indicate that there are other factors in the process of transfer from crop to worker or that DFR may not be the best determinant of exposure in situations where re-entry takes place several weeks after the last application. Therefore, it was hypothesized that total or solvent-extracted foliar residue (SFR) might better represent the ability to transfer to the worker instead of DFR.

Because parts of the model are used in the Dutch pesticide authorization procedure to estimate worker exposure during re-entry, a study was conducted to validate some of the aspects of the proposed model. Emphasis was put on the relationship between the applied amount of active ingredient and the resulting increase of DFR in relevant zones (crop heights), as well as determining factors (i.e., application techniques and crop density, or leaf volume index). In addition, the influence of re-entry time and crop density on transfer factors (calculated from levels of re-entry exposure and either DFR or SFR) was studied.

Material and methods

Study design

This study was conducted in seven commercial greenhouses for the cultivation of carnations. In four greenhouses, the same pesticide was applied to the same crop. Alternately, a high-volume application (spray volume > 600 L/ha) or a low-volume application (spray volume < 50 L/ha) was used. The increases in the dislodgeable foliar residue (DFR) and the solvent-extracted foliar residue (SFR) resulting from application were assessed. Crop volume was measured at the time of each application in order to detect possible changes of the crop density due to growth of the crop. In addition, in three other greenhouses a single application was monitored.

After a high-volume application, a decrease of the DFR was measured by leaf sampling (in two zones) on ten occasions over a period of 4 weeks. Dermal and respiratory exposure were measured when workers were harvesting the treated crop at several times during the same 4-week period after each high-volume application. The changes in DFR could then be compared to the changes in exposure. When using low-volume techniques, only changes in DFR were assessed.

Three pesticides having large variation in recommended dose rate were selected for this study. All were considered chemically stable and non-volatile. Model pesticides that were selected were thiophanate-methyl (CAS 23564-05-08; tradename Topsin M®;[*] 500 g a.i./L), methiocarb (CAS 2032-65-

---

[*] Registered trademark of Schering Aagrunol BV; Haaren, The Netherlands.

7; tradename Mesurol Vloeibaar®;* 500 g a.i./L), and abamectin (CAS 71751-41-2; tradename Vertimec®;** 18 g a.i./L). The recommended concentrations in the application mixture were, respectively, 140, 100, and 25 mL formulation per 100 L water for high-volume techniques, resulting in 70, 50, and 0.45 g a.i./100 L, respectively. For low-volume techniques, the same application rate (g a.i./ha) was used as during high-volume application on the same crop.

Sample collection for DFR and SFR

Stratified whole-leaf sampling was conducted prior to and after application during a 4-week sampling scheme and during re-entry. Each time, 24 leaves from crop at a designated height were collected in duplicate in a 500-mL polyethylene bottle. Leaves to determine the increase of DFR and half-lives were collected at two different zones (high and low). The high zone was defined to be approximately 3/4 of the total crop height, whereas the low zone was defined as 1/4 of the total crop height. Dislodgeable foliar residues are all expressed in mg active ingredient per square centimeter (one-sided leaf area). The increase of dislodgeable foliar residue due to application was calculated using the difference between the average DFR level from duplicate leaf samples of each zone just before and just after the application.

Prior to the collection of leaves, the working technique of each worker was observed in order to define the contact zones to the crop. If there was a great difference (>10 cm) between corresponding contact zones of workers (cutting and collecting), leaf sampling was performed for each worker (personal DFR). If not, leaf sampling was done in the average contact zones of the crop, as indicated above.

Leaf samples were stored at 4 to 7°C in the laboratory until analysis. After extraction and analysis, the leaf volume was measured by stereometric volumetry using a method described by Sherle (1970). A relationship between leaf volume and leaf area was determined for carnation leaves by measuring leaf area (one-sided) of leaf samples with different leaf volumes using a surface area meter (LI-COR 3100, LI-Cor, Inc., Nebraska).

Sampling for dermal exposure during re-entry

Dermal exposure was performed using pre-washed cotton gloves covering the hands and forearms (stretch-cotton, 200 g/m²; surface (one-sided) 370 cm²; J. van der Wee BV; Riel, The Netherlands). A pair of gloves was used for a maximum period of one hour in order to prevent breakthrough of the pesticide(s). A new pair was provided after each hour of harvesting, or earlier

---

* Registered trademark of Merck, Sharp and Dohme BV; Haarlem, The Netherlands.
** Registered trademark of Bayer BV; Mijdrecht, The Netherlands.

in case a glove was damaged. In order to estimate possible differences of exposure and/or transfer of dislodgeable residue between the left and right hand, left and right gloves were collected separately in 1-L polyethylene bottles. Gloves were stored at 4 to 7°C until analysis.

Sampling for respiratory exposure during re-entry

Respiratory exposure was measured during re-entry (harvesting) simultaneously with dermal exposure. Measurements were carried out using an IOM sampler (IOM, Negretti Automation, England). The sampling head was attached to a constant-flow air pump operating at 2 L/min (P2500 Air Sampling Pump, DuPont). The respiratory fraction was estimated according to the American Conference of Governmental Industrial Hygienists (ACGIH, 1985). The air filter contained a glass fiber esterase filter (25 mm, Type A/E, Gelman Sciences) when sampling methiocarb or a mixed cellulose filter (25 mm, 8-mm pore diameter, Millipore Corp.) when sampling thiophanate-methyl. Flows were adjusted before and checked after the sampling period with a pre-calibrated rotameter tube (Rota 1/4-600, Dr Hennig GmbH, Germany). Air-sampling filters were put into 50-mL polypropylene tubes (Greiner und Söhne GmbH, Germany) and stored at 4 to 7°C (mixed cellulose esterase filter for thiophanate-methyl) or –20°C (glass fiber filter for methiocarb).

Crop density estimates

Crop volume was estimated using a non-destructive procedure as described by Bierman et al. (1997). This method was based on stereological methods as used in pathological research (Gundersen et al., 1988). A random sample was taken to estimate the percentage of space in which the carnation crop was situated. Test-points of a number of crossings from the upper grid (a metal netting used to prevent carnation stems from breaking) were defined, while excluding the two most outside wires. When a piece of crop (leaf, stem, or flower) made contact with a crossing, this was counted as a "hit." From results of multiple investigations on more beddings (the more test-points used, the smaller the error in estimate), a leaf volume estimate can be calculated using the following formula:

$$LVI = (\Sigma H/(4 \times n\text{-Tests})) \times 100\% \qquad (6)$$

where:

LVI = leaf volume index (v/v%)
ΣH = sum of the number of hits from each investigation
n-Tests = total number of points tested in n investigations

Additional observations

Picking rate

At the end of each harvesting period, the number of flowers harvested was counted during sorting and bundling. The number of flowers per worker was determined by dividing the total number of flowers by the number of workers involved in harvesting. This number was corrected for the estimated percentage of harvested flowers in case there was great difference in picking rate between workers.

Application rate

Tank-mix samples were taken prior to each application to determine actual spray concentration. For high-volume applications, the remaining spray liquid was estimated. After completing the low-volume application, the volume of the remaining spray solution was measured and another tank sample was taken. Application rates were calculated from the spray volume, acreage, and concentration of the active ingredient in the tank sample.

Chemical analyses

Leaf samples

Dislodgeable foliar residues were obtained with the method described by Iwata et al. (1977). Leaves were extracted twice by shaking for 30 min with 200 mL of distilled water containing 8 drops of Triton-X100 solution at 200 strokes per min. The bottle containing the leaves was rinsed with 50 mL of distilled water, the leaves removed, and the bottle rinsed twice with 20 mL methanol. The total volume of the fractions collected was measured and recorded. Leaves were returned to the polyethylene bottle. After the DFR procedure, solvent-extracted foliar residue was determined by manually shaking leaves for 2 min with 50 mL methanol. The extracts of DFR samples containing methiocarb were adjusted to pH 4.5 with acetic acid immediately following the DFR procedure to avoid instability of methiocarb caused by the Triton in the extract. This was not necessary after the SFR procedure. Leaf samples were rinsed with water to remove methanol and returned to storage at 4 to 7°C until measurement of leaf volumes/surface area could be performed.

Gloves

For extraction of thiophanate-methyl, methanol (250 mL per glove) was added to the gloves, and, for extraction of methiocarb, 250 mL 60/40 (v/v%) methanol/water (per glove) was added to the gloves and adjusted to pH 4.5 with acetic acid. Gloves were then ultrasonicated for 10 min and extracted for 30 min by shaking at 200 strokes per min.

Filters

For extraction of the mixed cellulose esterase filters, 5 mL methanol (100%) was added. For the glass fiber filters, 5 mL 60/40 (v/v%) methanol/water was used. Filters were then ultrasonicated for 10 min and extracted for 30 min by shaking at 200 strokes per min.

Abamectin

Solutions of dislodgeable foliar residue and absolute foliar residue containing abamectin were extracted with *n*-hexane. The extraction efficiency of abamectin from solutions of dislodgeable foliar residue into *n*-hexane was 100%. Abamectin residues were converted into a fluorescent compound by derivatization with trifluoroacetic anhydride and subsequent hydrolysis of the derivative with ammonium hydroxide. This solution was quantified by reversed-phase, high-performance liquid chromatography (HPLC) and fluorescent detection according to the method described by Jongen et al. (1991). Recovery was > 95 %. The detection limit of the method was 1 mg/L for extracts of gloves and filters, and 0.25 mg/L for leaf samples. The "between days" coefficient of variation (CV) was < 5%.

Methiocarb

Extracts of all matrices were analyzed by reversed-phase HPLC using ultraviolet detection at a wavelength of 266 nm (Soekhoe and Kerstens, 1995). The limit of detection (LOD) was 10 mg/L for all matrices. Recovery was > 90% and "between days" CV of the analytical chemical method was < 10%.

Thiophanate-methyl

Extracts of gloves, extracts of foliar dislodgeable residue and absolute foliar residues, and filters containing thiophanate-methyl were analyzed by reversed-phase HPLC and ultraviolet detection at 254 nm (Engel, 1988). The LOD was 50 mg/L for extracts of filters and gloves, and 10 mg/L for leaf samples. Recovery was > 90%, and the "between days" CV of the analytical chemical method was < 5%.

Statistical analyses

The data were statistically analyzed using the SOLO Statistical System (BMDP Statistical Software, Inc., Los Angeles, CA) on a personal computer. Differences between groups were tested by the Mann-Whitney test or a paired *t*-test in cases where paired data sets were tested. Possible relationships were studied with (multiple) linear regression using least-square estimates.

Results

Dislodgeable foliar residue and solvent-extracted foliar residue

The solvent-extracted foliar residue (SFR) was less than 0.5% of the dislodge-able foliar residue (DFR) in all cases; therefore, only results concerning DFR are presented in Table 1. Possible differences in the rate of increase of DFR with application (i.e., ΔDFR) within the same crop and between the high and low zone were tested for each application technique using a paired *t*-test. For all high-volume applications, ΔDFR (adjusted to application rate) in the higher zone was significantly higher than the increase in the lower zone. In cases where low-volume equipment was used, no significant difference between ΔDFR for both crop heights was observed. This indicates that there might be a difference in distribution of residue to the crop for the two application techniques.

Table 1 Increase of DFR (ΔDFR) in Two Crop Zones (High and Low) and Average Increase with Application

Green-house	Appli-cation	Active ingredient	Applica-tion rate (g a.i./m^2)	ΔDFR$_{high}$ (μg a.i./cm^2)	ΔDFR$_{low}$ (μg a.i./cm^2)	ΔDFR$_{average}$ (μg a.i./cm^2)
1	HV	Tpm	0.150	3.93	2.59	3.25
2	HV	Tpm	0.108	2.81	0.51	1.66
2	LV	Tpm	0.062	0.66	0.42	0.54
3	HV	Tpm	0.172	4.05	2.28	3.17
3	LV	Tpm	0.076	2.23	1.03	1.63
4	HV	Tpm	0.053	2.99	1.35	2.18
4	HV	Meth	0.052	3.29	1.66	2.48
4	LV	Meth	0.028	0.25	0.45	0.35
5	HV	Meth	0.053	1.05	0.86	0.96
5	LV	Meth	0.036	0.76	0.49	0.63
6	HV	Meth	0.024	1.04	0.13	0.59
7	HV	Aba	0.001	0.0042	0.0033	0.0038

Note: HV = high-volume application; LV = Low-volume application; Tpm = thiophanate-methyl; Meth = methiocarb; Aba = abamectin.

Uniformity of ΔDFR between high and low crop zones vs. application technique was tested. A statistically significant difference was observed between both application techniques within the same greenhouse (n = 4; paired *t*-test, p <0.01 for average ΔDFR [mg/cm^2]). However, no significant differences were observed for the average ΔDFR (adjusted to applied application rate) between both application techniques within the same greenhouse. Because significant differences were observed between application rates for high-volume and low-volume applications within the same greenhouse (*t*-test, p < 0.01), the application rate seems to be a major determinant for an increase in average DFR with application.

Table 2 Relationships Between ΔDFR (mg/cm²) Averaged for
High and Low Crop Zones, and Application Rate (g/m²)

Technique	Intercept	Regression coefficient	R^2_{var}[a]	R^2_{model}[b]
High-volume	0.51	16.6	0.71	0.66
(n = 8)	p = 0.26	p = 0.009	p = 0.009	
Low-volume	–0.26	20.7	0.65	0.47
(n = 4)	p = 0.70	p = 0.20	p = 0.20	
Both	0.25	17.7	0.66	0.63
(n=12)	p = 0.46	p = 0.001	p = 0.001	

[a] Amount of variance explained by regression.
[b] Total amount of variance explained by regression, adjusted for sample size (n).

The relationship between average ΔDFR and application rate (AR) was tested using linear regression analysis and the model ΔDFR = a + b AR. All data concerning high-volume applications (n = 8) and low-volume applications (n = 4) were used, separately or together. The results of these regression analyses are presented in Table 2.

For the low-volume technique, no significant relationships were observed. For the high-volume technique and high- and low-volume techniques together, a significant relationship was observed. The high-volume technique and the low-volume/high-volume technique showed a variation of 66% and 63%, respectively. The relationship of increase of average DFR and application rate is illustrated in Figure 1.

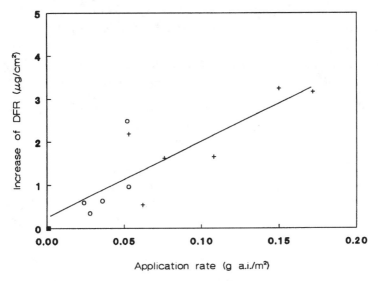

Figure 1 Increase of average DFR of methiocarb (o), thiophanate-methyl (+), and abamectin (■) due to application with either high-volume or low-volume techniques.

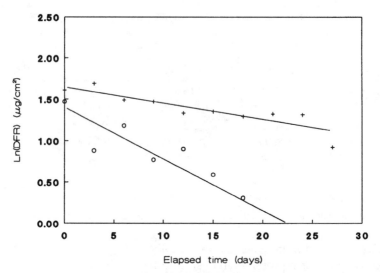

Figure 2 Typical log-transformed decays of DFR for estimating half-lives of methiocarb (o) and thiophanate-methyl (+).

Crop volume estimates ranged from 6.72 to 17.1 (v/v%), with a median
of 11.1 and a mean of 11.3 (v/v%). The differences between crop volume
estimates at the moment of HV and LV applications ranged from 0 to 3.8
(v/v%), depending on the elapsed time between applications.

A possible relationship between DFR and the application rate, as well
as the crop volume estimate (CrV), was investigated using a multiple linear
regression model ($\Delta DFR = a + b\,AR + c\,CrV$). No significant contribution of
crop volume to the variation of ΔDFR was observed ($p = 0.19$ and $p = 0.87$
for high-volume applications and all applications, respectively).

Dislodgeable foliar residue as a function of time and estimating half-lives of the pesticides after high-volume application

The DFR values were followed over a period of 4 weeks from the high-
volume application. The decrease of DFR in the two zones was monitored
in six greenhouses (three times after application of thiophanate-methyl and
three times after application of methiocarb). Figure 2 shows typical log-
transformed decays of DFR (average of two samples for each zone) for
methiocarb and thiophanate-methyl. Assuming a first-order decay, half-lives
were calculated using Equations (3) and (4) and were found to be 29 ± 8.5
days and 11 ± 3.4 days for thiophanate-methyl and methiocarb, respectively.

The relationship between exposure and determinants of exposure

The relationship between dermal and respiratory exposure during harvesting of carnations and the level of DFR on the crop was investigated in six

Table 3 Range, Median, GM, and GSD of Assessed Total Dermal and
Respiratory Exposure to Thiophanate-Methyl and Methiocarb

	Thiophanate-methyl	Methiocarb
Dermal	4.56–31.5 (8.24)[a]	0.23–5.70 (1.61)[a]
exposure	9.19 (1.6)[b]	1.34 (2.5)[b]
(mg/hr)		
Respiratory	8.79–57.6 (21.5)[a]	0.2–8.1 (1.14)[a]
exposure	20.4 (1.6)[b]	0.99 (3.3)[b]
($\mu g/m^3$)		

[a] Range (median).

[b] GM (GSD). GM = geometric mean; GSD = geometric standard deviation.

greenhouses. At each greenhouse, two workers were monitored during 2 to 5 consecutive re-entries within a 4-week period after the last high-volume application. Descriptives (i.e., range, median, and mean) of dermal and respiratory exposure are presented in Table 3. Full details on exposure data are described elsewhere (de Haan et al., 1996). Both dermal and respiratory exposure to thiophanate-methyl exceed exposure to methiocarb substantially. The average picking rate differed between harvested flower types after application of thiophanate-methyl (616 flowers, range = 417 to 1056) and after application of methiocarb (mean = 435 flowers, range = 276 to 784). Re-entry times ranged from 1 to 26 days after high-volume application.

Dermal exposure

Regression analyses were performed to investigate relationships between dermal exposure and possible determinants of re-entry exposure — average DFR of the contact zones, number of flowers harvested per hour, and re-entry time (days after application) (Tables 4 and 5). Because there was a large influence on total regression analyses for one data set for one worker, this data set was determined to be an outlier and excluded from further analyses. Possible correlations between the three independent variables were tested and observed as r ≤ 0.19. All variables were shown to contribute significantly to the explained variation ($R^2 = 0.76$). A transfer factor (TF) of 2262 cm^2/hr was estimated. When only data involving exposure to thiophanate-methyl were used, no relationship ($R^2 = 0.007$, $p = 0.75$) was observed between total dermal exposure (both hands) and average DFR. Because DFR is strongly correlated with number of flowers and re-entry time (correlations are r = 0.68 and r = 0.45, respectively), the only relationships found were between total dermal exposure and number of harvested flowers as well as between total dermal exposure and re-entry time (days after application). When all data on thiophanate-methyl were excluded from the analyses, both the significance of all variables and the total explained variation of the regression model increased ($R^2 = 0.89$).

Table 4 Regression Analyses of All Total Dermal Exposure Data
(mg/hr) as Dependent Variable to Average DFR,
Number of Harvested Flowers, and Re-entry Time

Variable	Intercept estimate[a]	Regression coefficient[a]	R^2_{var}[b]	p-value	R^2_{seq}[c]
Average DFR ($\mu g/cm^2$)		2262 (292)	0.53	<0.0001	0.53
Number of flowers (per hr)		13.3 (2.09)	0.33	<0.0001	0.74
Re-entry time (days)		–127 (45.9)	0.09	0.008	0.78
Model	–3611 (1283)			<0.0005	0.76[d]

[a] Standard error in parentheses.

[b] Amount of variance explaianed by variable.

[c] Total amount of variance explained by 1, 2, or 3 variables, respectively.

[d] Total amount of variance explained by multiple regression model, adjusted for sample size.

Table 5 Regression Analyses of Total Dermal Exposures to
Thiophanate-Methyl (mg/hr) as Dependent Variable
to Number of Harvested Flowers and Re-entry Time

Variable	Intercept estimate[a]	Regression coefficient[a]	R^2_{var}[b]	p-value	R^2_{seq}[c]
Number of flowers (per hr)		13.5 (3.2)	0.29	0.0009	0.29
Re-entry time (days)		–375 (64)	0.53	<0.0001	0.79
Model	7184 (2197)	<0.0005	0.76[d]		

[a] Standard error in parentheses.

[b] Amount of variance explained by variable.

[c] Total amount of variance explained by 1 or 2 variables, respectively.

[d] Total amount of variance explained by multiple regression model, adjusted for sample size.

To obtain evidence of a possible change of the transfer factor (TF) with time, the ratios of total dermal exposure and average DFR during re-entry period were calculated for each set of re-entry data. For each worker, regression analyses were performed for calculated ratios depending on re-entry time (days after application). For three out of ten persons investigated, the TF decreased significantly with an increase of re-entry time. For all other workers, a trend of TF decrease was observed. When regression analyses was performed on all calculated transfer factors (depending on re-entry time), a significant decrease of TF ($R^2 = 0.11$, $p = 0.025$) was observed once more.

Respiratory exposure

Eight out of 23 observations of the respiratory exposure to methiocarb were below the level of quantification (LOQ) of the analytical chemical method; therefore, upper limits of possible respiratory exposure levels were calculated using the amount reflecting 1/2 LOQ and duration of exposure (ranging from <0.12 to <0.18 mg/m³). The level of respiratory exposure to thiophanate-methyl was generally higher than the respiratory exposure to methiocarb. Regression analyses were performed for each pesticide separately. Regression analyses for methiocarb were only performed on exposure levels > LOQ. Because crop volumes were almost equal for the remaining data, the influence of crop volume as an independent variable of respiratory exposure could not be analyzed. Strong correlations were observed between re-entry time and the number of flowers harvested per hour ($r = 0.36$ and $r = 0.58$ for thiophanate-methyl and methiocarb, respectively). Respiratory exposure levels were first corrected for the number of flowers harvested, assuming that respiratory exposure is the result of resuspension of organic foliar dust. The number of harvested flowers per hour was a measure for the frequency of crop disturbance. Significant relationships were observed in linear regression analyses with respiratory exposure ([mg/m³]/[flowers/hr]) as a dependent variable and average DFR as an independent variable for both pesticides ($R^2 = 0.43$, $p = 0.008$ for thiophanate-methyl; $R^2 = 068$, $p = 0.006$ for methiocarb).

From the regression analyses it is clear that significant relationships were observed between respiratory exposure to thiophanate-methyl and methiocarb (adjusted to the number of flowers harvested) and dislodgeable foliar residues during re-entry.

Discussion

Based on all application data, the results of this study indicate a linear relationship between the increase of the dislodgeable foliar residue and application rate; however, for the low-volume applications, this relationship is not statistically significant. The relatively limited number of data may be the reason for this.

Within greenhouses, no significant differences were observed between ΔDFR adjusted to the application rate for both application techniques. When ΔDFR values were not adjusted to actual application rates, the differences were statistically significant, as the application rates of low-volume techniques were lower than those of high-volume techniques. Comparison of the concentration of the active ingredient in the tank mix showed higher concentrations at the end of the application than at the start, indicating a non-uniform mixture of the formulation in the tank. Moreover, in most cases, a considerable volume of highly concentrated spray liquid remained in the tank due to the construction of the suck-pipe inside the tank.

The results of the DFR assessment of different crop zones indicate that low-volume applications result in a more homogeneous distribution over the crop compared to high-volume applications. A recent study on the interception of high-volume applications in the cultivation of chrysanthemums revealed interception ratios from 0.2 to 1 related to the leaf area index (LAI) (Veerman et al., 1994). In our study, it was not easy to assess the LAI because of the structure of the carnation crop. Estimation of the LAI based on the results of estimation of the crop density (leaf volume index) was not reliable enough and resulted in a large variance of the calculated interception ratio (from 0.4 up to 5).

Data on the DFR of chlorothalonil on carnation crops in previous studies indicated an increase in DFR of 10 to 20% with solvent extraction (unpublished data). However, in all of the current experiments, no differences were observed between dislodgeable foliar residue and solvent-extracted foliar residue for the pesticides methiocarb and thiophanate-methyl.

The dermal exposure during re-entry decreased by factors of 2 to 11 during the observed 4-week period after application, largely due to the decrease of the DFR during that period. Decay of the pesticide, on the one hand, and dilution (e.g., growth of the foliage and moving up of the worker-crop contact zone), on the other hand, may be responsible for this decrease. In addition, the decrease of the picking rate contributed to the decrease of dermal exposure.

The level of dermal exposure is significantly associated with re-entry time ($R^2 = 0.09$, $p = 0.008$). Primarily, the decrease of DFR in time will be responsible for the decrease of dermal exposure. There may also be a decrease in the transfer factor due to an increase in re-entry time (Veerman et al., 1994). At the end of the observed 4-week re-entry period, a decrease in the TF of 7.6% of the initial TF was observed using all dermal exposure data. The decrease of the TF at the end over the same period was 3.4% when using only dermal exposure data of methiocarb. A significant decrease in the transfer factor ($R^2 = 0.11$, $p = 0.025$) over time was also observed when regression analyses were performed on all calculated transfer factors (ratios of total dermal exposures and corresponding average DFRs), depending on re-entry time. No differences were observed over the period of observation between SFR and DFR. A suspected change of dislodgeability was not shown.

The average value of the transfer factor (approximately 2300 cm^2/hr) derived from all data in our study differs somewhat from the transfer factor of 3300 cm^2/hr observed in a preceding study during the cultivation of carnations (Brouwer et al., 1992). However, that transfer factor had been derived from data on both high-volume spray (as in the present study) and dusting operations. The powder formulations showed an effective transfer of the residue to the body. Recalculated transfer factors for only high-volume applications showed an average of approximately 2500 cm^2/hr.

The explained variation of the dermal exposure in the present study by the transfer factor is relatively low ($R^2 = 0.53$) compared to the $R^2 = 0.66$

observed in a previous study (Brouwer et al., 1992). As in the present study, thiophanate-methyl residue was shown to result in strongly different transfer factors compared to other pesticides. The average transfer factor of about 1600 cm^2/hr observed for methiocarb ($R^2 = 0.70$) corresponds with the results of an intervention study in the cultivation of carnations where an average transfer factor of 1400 ($R^2 = 0.87$) was observed (Emmen et al., 1996).

Popendorf et al. (1975) suggested a relationship between respiratory exposure and organic foliar dust released from the foliage due to crop disturbance; therefore, it was hypothesized that crop density may be associated with both dermal and respiratory exposure. However, in the present study, no such relationship could be observed. This may be due to the large variation in the method used to determine crop volumes. Improvement of the crop-volume/leaf-surface-area method may contribute to the clarification of whether crop density can be considered a determinant of re-entry exposure.

A relationship between respiratory exposure (adjusted to the number of flowers harvested) and average DFR was observed for both pesticides. This indicates that respiratory exposure is the result of resuspension of organic foliar dust containing pesticide residue. The picking rate was used as a measure for the frequency of crop disturbance. For both pesticides, significant relationships were found ($R^2 = 0.43$, $p = 0.008$ for thiophanate-methyl; $R^2 = 0.68$, $p = 0.006$ for methiocarb) between adjusted respiratory exposure and average DFR. The regression coefficient of the relationship is expressed in cm^2/hr/m^3/flower and amounted to 0.017 for thiophanate-methyl and 0.0048 for methiocarb. This regression coefficient can be considered to be comparable with the TF for dermal exposure. Conceptually, it equals the fraction of the foliar residue that can be resuspended in 1 m^3 air during harvesting of a certain number of flowers per hour. Using the average picking rates for both pesticides, as observed in this study, these resuspension factors (RFs) can be adjusted to 2.1 cm^2/m^3 and 10.5 cm^2/m^3 for methiocarb and thiophanate-methyl, respectively. The large difference between picking rates may be explained by influence of the season on the use of both pesticides (insecticide vs. fungicide) and the decrease in the number of harvestable flowers per square meter of crop during the period of use of the fungicide.

Half-life estimates of approximately 28 days for thiophanate-methyl indicate a very slow decay compared to methiocarb with an estimate of half-life of about 11 days. The application of a model based on a first-order decay process resulted in fairly high R^2 and significant fit. The results suggest that both pesticides are relatively stable compared to other compounds under similar environmental conditions (Brouwer et al., 1994). With respect to the objectives of the study and the proposed model, it can be stated that the results confirm the assumption of a linear relationship between application rate (for both application techniques) and the increase of dislodgeable foliar residue. This relationship holds for modeling purposes. The contribution of the crop density or total crop surface area to the process of interception cannot be quantified with the results of the present study. Because the interception factor ranges from about 0.35 to 0.9 (Willis and McDowell, 1987), the error of the estimate of DFR will be limited to a factor of 3. However, an estimate of crop or foliar surface area is also necessary to calculate the increase of DFR from the application rate. This emphasizes the necessity of having relatively simple methods that can be used for a wide range of crops and result in reliable estimates of crop surface area.

Because half-lives may vary substantially, the values of the actual DFR available for transfer to the worker will diverge with prolonged re-entry time. For example, the difference between the actual DFRs will increase approximately 3% per day for both pesticides in our study. These pesticides have long half-lives which differ by a factor of 2, assuming a similar initial residue deposit; therefore, the half-life of a pesticide must be considered a critical factor for the level of re-entry exposure. Increase of re-entry time will decrease the transfer factor; however, the decrease observed in this study is small and negligible compared to all other factors and variances that influence re-entry exposure.

Harvesting carnations will account for average transfer factors of about 2000 cm²/hr for pesticides applied by high-volume applications. A critical evaluation of existing data may generate more reliable generic transfer factors for use in the re-entry model.

In conclusion it can be stated that the basic assumptions of the re-entry model — a linear relationship between application rate and initial dislodgeable foliar residue and a first-order decay of the DFR — have been confirmed. The relationship between the transfer factor and re-entry time at various DFR levels should be explored further. Including information on foliage surface area or crop density may lead to a refinement of the model; however, crop volume estimating methods should be improved before their influence on the exposure processes can be fully evaluated.

Acknowledgments

The authors wish to thank the Ministry of Social Affairs and Employment for the financial support for a series of research projects to investigate worker exposure to pesticides. We want to thank Sjaak de Vreede, Bert Bierman, John Matulessy, and Ceciel Lansink for their assistance during the field work. For performing all chemical analyses, we extend special thanks to Roel Engel, Usha Soekhoe, and Lambert Leenheers. Finally, all growers of carnations who participated in this study, and without whom it would not have been possible to perform this investigation, are gratefully acknowledged.

References

ACGIH (1985) *Particle Size Selective Sampling in the Workplace*, American Conference of Governmental Industrial Hygienists, Cincinnati, OH.

Bates, J.A.R. (1990) The prediction of pesticide residues in crops by the optimum use of existing data, *Pure/Appl. Chem.*, 62:337–350.

Bierman, E.P.B, Brouwer, D.H., and van Hemmen, J.J. (1997) Measuring crop density: comparison of volumetry and stereological methods, *Bull. Environ. Contam. Tox.*, 58:1006–1013.

Brouwer, D.H., de Haan, M., Peelen, S., van de Vijver, L., and van Hemmen, J.J. (1994) Dislodgeable foliar residue as an estimate of source strength for worker exposure to pesticides, in *Book of Abstracts from the 8th International Congress of Pesticide Chemistry*, American Chemical Society/IUPAC, July 4–9.

Brouwer, R., Brouwer, D.H., De Mik, G., and van Hemmen, J.J. (1992) *Exposure to Pesticides*. Part I. *The Cultivation of Carnations in Greenhouses*, S131-1, Ministry of Social Affairs and Employment, The Hague, The Netherlands; also published in *Am. Ind. Hygiene Assoc. J.*, 53:575–581, 1992.

Emmen, H.H., Hoogendijk, E.M.H., Brouwer, D.H., Muijser, H., and Kulig, B.M. (1996) *Cumulative Effects of Pesticide Exposure on Human Nervous System Functioning*, Report V96.320, TNO Nutrition and Food Research Institute, Zeist, The Netherlands.

Engel, R. (1988) *Determination of Thiophanate-methyl by High Performance Liquid Chromatography and UV Detection*, Report BT-41, Medical Biological Laboratory TNO, Department of Occupational Toxicology, Rijswijk, The Netherlands (in Dutch).

Gundersen, H.J.G., Bendtsen, T.F., Korbo, L., Marcussen, N., Moller, A., Nielsen, K., Nyengaard, J.R., Pakkenberg, B., Sorensen, F.B., Vesterby, A., and West, M.J. (1988) Some new, simple and efficient stereological methods and their use in pathological research and diagnosis, *APMIS*, 96:379–394.

Haan de, M., Brouwer, D.H., and van Hemmen, J.J. (1996) *Re-entry Exposure Estimates: Application Technique, Foliar Surface and Re-entry Time as Critical Predictors for Dislodgeable Foliar Residue*, Report V96.384, TNO Nutrition and Food Research Institute, Zeist, The Netherlands.

Iwata, Y., Knaak, J.B., Spear, R.C., and Foster, R.J. (1977) Worker re-entry into pesticide-treated crops. I. Procedure for determination of dislodgeable foliar residues on foliage, *Bull. Environ. Contam. Tox.*, 10:649–655.

Lundehn, J.R., Westphal, D., Kieczka, H., Krebs, B., Löcher-Bolz, S., Maasfeld, W., and Pick, E.P. (1992) *Mitteilungen aus der Biologische Bundesanstalt für Land- und Forstwirtschaft Bundesrepublik Deutschland, Einheitliche Grundsätze zur Sicherung des gesundheitsschutzes für den Anwender von Pflanzenschutzmitteln*, Heft 277, Berlin, Germany.

MAFF/Joint Medical Panel of the Scientific Subcommittee on Pesticides of the Ministry of Agriculture, Fisheries, and Food and the Toxicology Committee of the British Agrochemical Association (1986/1992) *Estimation of Exposure and Absorption of Pesticides by Spraying Operators*, MAFF, Pesticides Registration Department, Harpenden Laboratory, Harpenden, Herts, England.

Jongen, M.J.M, Engel, R., and Leenheers, L.H. (1991) High performance liquid chromatography method for the determination of occupational exposure to the pesticide abamectin, *Am. Ind. Hygiene Assoc. J.*, 52: 433–437.

Popendorf, W.J. (1985) Advances in the unified model for re-entry hazards in Dermal exposure related to pesticide in use, in *Discussion of Risk Assessment*, Honeycutt, R.C., Zweig, G., and Ragsdale, N.N., Eds., ACS Symposium Series 273, American Chemical Society, Washington, D.C., pp. 323–340.

Popendorf, W.J., Spear, R.C., and Selvin, S. (1975) Collecting foliar pesticides residues related to potential airborne exposure to workers, *Environ. Sci. Technol.*, 9:583–585.

Sherle, W. (1970) A simple method for volumetry of organs in quantitative stereology, *Microscopy*, 20:57–60.

Soekhoe, U. and Kerstens, H.J. (1995) *Determination of Propoxur, Methiocarb, and Tinopal on Gloves, Filters, and Leaves by High Performance Liquid Chromatography*, Report DATV/BT/076, TNO Nutrition and Food Research Institute, Department of Occupational Toxicology, Rijswijk, The Netherlands (in Dutch).

van Hemmen, J.J. (1992) *Assessment of Occupational Exposure to Pesticides in Agriculture:* Part I. *General Aspects*, Part II. *Mixing and Loading*, Part III. *Application*, S-reeks S141 1/3, Ministry of Social Affairs and Employment, The Hague, The Netherlands; also, Agricultural pesticide exposure data bases for risk assessment, *Rev. Environ. Contam. Toxicol.*, 126:1–85, 1992.

van Hemmen, J.J. (1993) Predictive exposure modelling for pesticide registration purposes, *Ann. Occup. Hygiene*, 37:514–564.

van Hemmen, J.J., van Golstei, Y.G.C., and Brouwer, D.H. (1995) Pesticide exposure and re-entry in agriculture, in *Methods of Pesticide Exposure Assessment*, Maroni, M., Iyengar, S., Curry, P., and Maloney, P., Eds., Plenum Press, London, pp. 9–19.

Veerman, M.C., van de Vijver, L., de Haan, M., Brouwer, D.H., and van Hemmen, J.J. (1994) *Exposure to Pesticides*. Part IV. *The Harvesting of Chrysanthemums in Greenhouses*, Report S131-5, Ministry of Social Affairs and Employment, The Hague, The Netherlands.

Willis, G.H. and McDowell, L.L. (1987) Pesticide persistence on foliage, *Rev. Environ. Contam. Toxicol.*, 100:23–62.

chapter ten

A foliar dislodgeable study of Curalan® DF in turf

S.C. Artz, J.R. Clark, and R.S. Kludas

Contents

Abstract

A foliar dislodgeable study in turf using Curalan® DF was performed to help in the determination of the risk of post-application exposure to humans. The study was performed in California, Florida, and Pennsylvania. Four applications of Curalan® DF were made at a target rate of 5.6 lb of active ingredient per acre. This was equivalent to one half the maximum label rate, which approximates the rate most often used by growers. The applications were made with a tractor-mounted boom sprayer at about 80 gallons of finished spray per acre. Two techniques were used to determine the amount of residue that could potentially be available to exposure: cloth dosimeters and turf clippings. Both cloth dosimeters and turf clippings were collected before and after each of four applications and then approximately 1, 2, 3, 7, 10, 14, 21, 28, 35, 42, 49, 56, and 63 days after the last application. In general, residues in both matrices at all three sites declined rapidly after the final application. Data from the turf clipping technique showed that the Dt_{50} (time when 50% of the initially applied residue is unavailable for physical removal from the treated area) ranged from 1 day in Florida to 5 days in Pennsylvania. The data from the California roller technique resulted in a Dt_{50} that ranged from 10 hr in Florida to 2 days in Pennsylvania. The Dt_{90} (time when 90% of the initially applied residue is unavailable for physical removal from the treated area) ranged from 2 days in Florida to 9 days in Pennsylvania. The coefficient of determination (r^2) values for the Dt_{50} curves were in the 0.9s, showing excellent correlation of the data to the Gustafson model (Gustafson and Holden, 1990).

Introduction

The ability to accurately determine the amount of exposure of a person to a chemical applied, in this case, to turf has been strongly pursued by those responsible for making judgments as to the amount of risk that is considered acceptable for the end user. Many different foliar dislodging techniques have been used as tools to estimate exposure to a compound of interest in the determination of transfer coefficients on turf. No one technique, thus far, has been determined to give what is considered the "correct" amount of residue dislodged during activity on treated turf. Such methods as the turf-clipping technique, the California roller technique, the PUF roller technique, the shuffle technique, and the drag-sled technique are currently being used throughout the pesticide industry to generate transfer coefficients for chemicals applied to turf. None of these, however, is more generally accepted than another at this time. The Outdoor Residential Exposure Task Force (ORETF) is currently involved in determining which technique provides the best and most consistent estimate of exposure to a chemical on turf.

The turf-clipping technique, which was used in the generation of some of the data that will be presented in this paper, consists of clipping the grass, adding a detergent solution to the clippings, and shaking the clippings for a long enough period of time to physically remove the chemical residues that are on the turf. The dislodging solution is then analyzed for residues.

The California roller technique, which also was used to generate some of the data in this study, consists of placing a defined-size piece of cloth material onto the treated turf, covering the cloth with a plastic sheet, and then rolling over the plastic and cloth sheets with a roller that contains 25 lb of lead shot. Typically, the roller is pushed back and forth over the sheets 10 times. The cloth is then extracted and analyzed for residues.

The PUF roller technique consists of placing a cylinder of polyurethane foam (PUF) onto a weighted, stainless steel roller and then rolling it over a defined area of treated turf. The turf is typically rolled over three times. The PUF is then pulled off the weighted, stainless steel roller, extracted, and analyzed for residues.

The shuffle technique consists of placing a cloth dosimeter onto the bottom of two separate platforms that are slightly bigger than a human foot. A platform is attached securely to each one of the feet of the investigator and the investigator then "shuffles" over a prescribed area of treated turf two times. The cloth is removed from each platform, extracted, and analyzed for residues.

The other popular technique in use today is the drag-sled technique, which consists of attaching a cloth dosimeter onto the bottom of a square block of some material (such as aluminum) of a predetermined size which has a weight added to the top side. The weighted sled is then pulled one time over a prescribed area of treated turf. The cloth dosimeter is taken off the sled, extracted, and analyzed for residues.

Each of these dislodging techniques tries to mimic ways that a person of a certain size or weight could contact the residues applied to turf. This is important because, should one of the techniques provide a more consistent means of dislodging than the others, it would allow risk assessments to be done that would give a good indication of the potential exposure and thus risk to an individual who comes in contact with the treated turf. It would also allow for better estimates to be made as to when the treated turf could be used by a typical homeowner.

In the present study, the results from the turf-clipping technique and the California roller technique will be contrasted and discussed. It is hoped that this will give the reader a better understanding of some of the difficulties encountered in determining the risk associated with exposure of chemical residues on turf by humans.

Materials and methods

Curalan® DF is a fungicide produced by BASF Corp. for the purpose of treating such diseases as brown patch and fusarium patch on turf. Curalan® DF was applied to turf in California, Florida, and Pennsylvania at a rate of 5.6 lb of active ingredient in 80 gallons of finished spray solution per acre using a tractor-mounted boom sprayer. Four applications were made 14 days apart. These conditions were selected to reflect the maximum level of dislodgeable residue when the product is typically used in agricultural practice.

The site in Madera, CA, consisted of a tall fescue/rye mix. The Florida site, located in Keystone Heights, consisted of Bermuda grass. The Pennsylvania site, in Germansville, consisted of bluegrass. Thus, this study was performed on three major grass varieties in three distinctly different temperature zones in the U.S.

Samples were taken before and after each of the four applications and at 1, 2, 3, 7, 10, 14, 21, 28, 35, 42, 49, 56, and 63 days after the fourth and last application. Field fortifications, in triplicate, were taken at each application and at 10, 35, 49, and 46 days after the last application in California; after each application and at 10, 35, and 49 days after the last application in Florida; and at 10, 35, 49, 46, and 63 days after the last application in Pennsylvania.

The two techniques used for determining the amount of residue available for human exposure were the turf-clipping technique and the California roller technique. The residues from the turf clippings were obtained from a

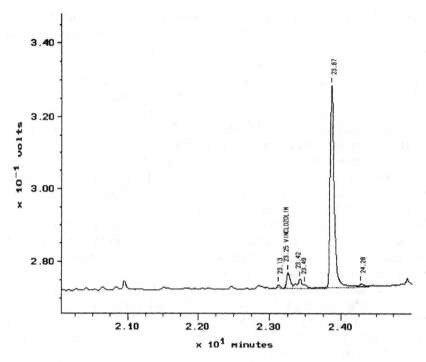

Figure 1 Control chromatogram from turf clipping.

10-g subsample of the clippings that were taken from a 40-cm × 40-cm area of treated turf where the grass was clipped all the way to thatch level. The clippings were immediately placed in a cooler and then dislodged off site. The residues were dislodged from the clippings by adding 150 mL of 0.01% Aerosol OT® (AOT), a surfactant, to the sample jar containing the clippings. The jar was shaken for 10 min and then the AOT was decanted into another clean sample jar. This procedure was repeated with another 150 mL of AOT, and then the AOT was added to the other 150 mL of AOT from the first dislodging. This was a physical dislodging of surface residue from the leaves, not a chemical extraction.

The extraction of the AOT samples consisted of taking a 50-mL subsample and extracting by separatory funnel with hexane and 10% NaCl in water. The cleaned-up and dried hexane extract was then injected onto a gas chromatograph (GC) with an ECD (electron conductivity detector). The minimum quantifiable limit (MQL), or limit of quantification, for the method was 0.002 µg/mL dislodging solution. A typical set of chromatograms can be seen in Figures 1 and 2.

The cloth dosimeter from the California roller technique consisted of placing a 15.2-cm × 61.0-cm piece of cotton cloth on treated turf, covering the cloth with plastic, and rolling over it with a 25-lb roller. The cloth was then placed in a sample container, which was put in a cooler and transported to the laboratory. The dosimeter was then extracted with hexane. An aliquot of

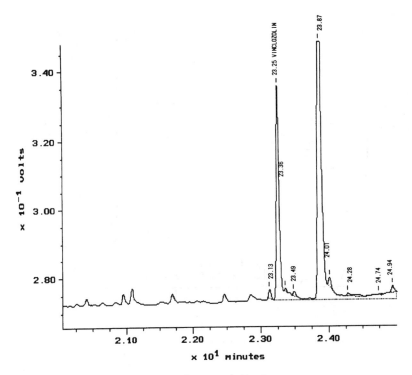

Figure 2 Sample chromatogram from turf clipping.

the extract was taken and injected onto a GC/ECD. The MQL for the method was 1.0 µg/cloth dosimeter. A typical set of chromatograms can be seen in Figures 3 and 4. All samples were taken in triplicate for both techniques.

Results

The results from all three sites show typical decline curves for the residues over time. In many cases, the declines are not linear but biphasic (Figures 5 through 10). They tend to form curves which are more accurately modeled using the Gustafson model. As noted on the curves in Figures 5 through 10, the correlation of the curves to the data is excellent. The coefficients of determination (r^2) for the curves ranged from the high 0.8s to the high 0.9s; thus, the data are accurately described by this model. Typically, Dt_{50} values are used to better describe the time when 50% of the initially applied residue is unavailable for physical removal from the treated area. When the model was run, the equation used to calculate the amount of residue that was unavailable was

$$Dt_{50} = (0.5^{-(1/@2)} - 1)/@3$$

where @2 and @3 are coefficients in the calculated equation for the fitted line.

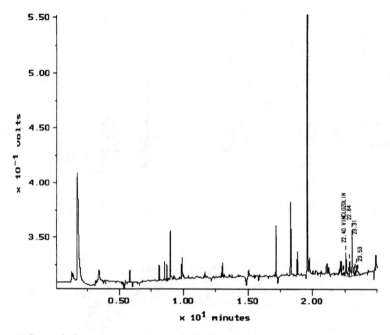

Figure 3 Control chromatogram from cloth dosimeter.

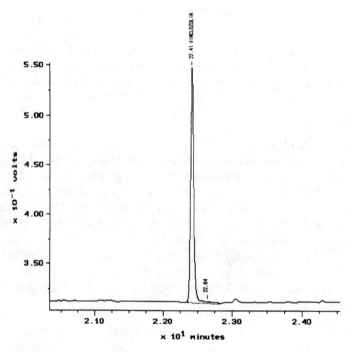

Figure 4 Sample chromatogram from cloth dosimeter.

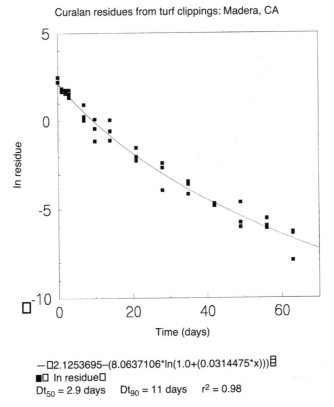

Curalan residues from turf clippings: Madera, CA

$$-\square 2.1253695-(8.0637106*\ln(1.0+(0.0314475*x)))\boxminus$$
■□ ln residue□
$Dt_{50} = 2.9$ days $Dt_{90} = 11$ days $r^2 = 0.98$

Figure 5 Decline curve for the California site using the turf-clipping method.

A similar equation is used to give an estimation of the time when 90% of the initial leaf residue is unavailable for physical removal and is represented by the equation:

$$Dt_{90} = (0.1^{-(1/@2)} - 1)/@3$$

Table 1 gives the half-lives, Dt_{50} and Dt_{90}, for the results from the three experimental sites. Table 2 contains the values for the amount of residue found at 0 day after the fourth application.

Discussion

The purpose of generating the numbers in Tables 1 and 2 is to ultimately determine a transfer coefficient. The transfer coefficient is used to calculate the exposure and finally a re-entry health risk to a chemical applied, in this case, to turf. Transfer coefficients are also calculated for other agricultural commodities to determine an estimate of risk when working in a treated crop, but this is not a topic of discussion for this paper.

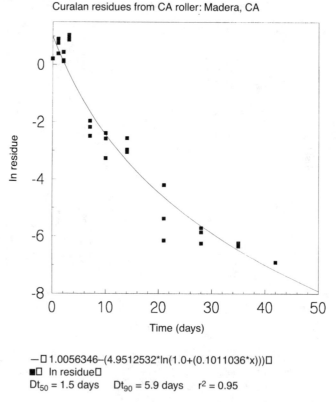

Curalan residues from CA roller: Madera, CA

—☐ 1.0056346–(4.9512532*ln(1.0+(0.1011036*x)))☐
■☐ ln residue☐
$Dt_{50} = 1.5$ days $Dt_{90} = 5.9$ days $r^2 = 0.95$

Figure 6 Decline curve for the California site using the California roller method.

A transfer coefficient (TC) is calculated in the following manner:

$$TC = D_{pot}/C_{envir}$$

where D_{pot} = the potential human dermal dose generated using passive dosimetry, and C_{envir} = the concentration of residue in an environmental matrix like turf foliar dislodgeable residue.

When the residues are dislodged from turf, the C_{envir} term can be determined, as the amount of residue per unit area that was sampled can be determined; therefore, the units of the C_{envir} term are typically $\mu g/cm^2$. To calculate a transfer coefficient, however, additional work must be done to determine the amount of residue recovered on someone who performs a certain task on turf. This could be exercising, playing ball, running, rolling, etc.

Because the transfer coefficient is calculated based on residues detected from the performance of a task, the final number is task dependent. The transfer coefficient is also dependent on the way the residues are dislodged from the turf, with different transfer coefficients being the result. Some of

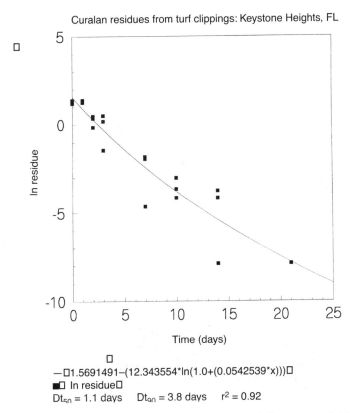

Figure 7 Decline curve for the Florida site using the turf-clipping method.

the things that can affect the measured dislodgeable residues on turf are weathering of residue, height of turf, the evenness with which the weight is applied to the turf being sampled, interaction of the applied chemical with the turf, formulation of the chemical, wetness of the turf, and the ability of the dislodging technique to physically remove dried chemical residue.

The results from this study were generated in three very different regions of the U.S. In the Madera region of California, the weather tended to be hot and dry during the study and residues tended to linger for longer periods of time. In the Florida region, rainfall was very prevalent, while the site in Pennsylvania tended to be on the cool and rainy side; residues there could linger unless there was rainfall.

Comparing the decline curves for both techniques, the results showed that the Dt_{50} times would be greatest in Pennsylvania (4.9 days), followed by California (2.9 days), with the least in Florida (1.1 days). It should also be noted that the ratios between the Dt_{50} and the Dt_{90} within each technique (Table 1) are very consistent, which indicates that the ability to physically dislodge the compound from the leaf surface was similar at each of the sites during the duration of the sampling and that the rate at which the

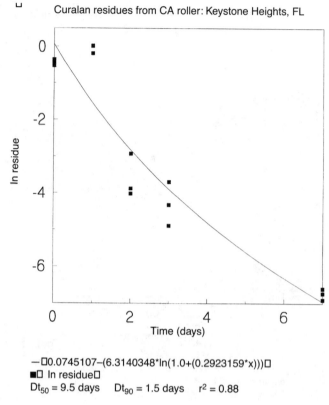

−□0.0745107−(6.3140348*ln(1.0+(0.2923159*x)))□
■□ ln residue□
$Dt_{50} = 9.5$ days $Dt_{90} = 1.5$ days $r^2 = 0.88$

Figure 8 Decline curve for the Florida site using the California roller method.

ability to physically dislodge this residue declined was very comparable from site to site and from technique to technique. Considering the non-reproducible nature of field studies and given that these two techniques were performed by different investigators in different parts of the country with different weather conditions, these decline curves and Dt measurements are similar.

As seen in Table 2, the results obtained from the two dislodging techniques, turf clipping and California roller, were different. It would be expected that the turf-clipping technique would give the highest residue results as the turf is actually clipped and not just rubbed as with the California roller technique. With the California technique, residues that are caught on the side of the leaf that was bent down toward the ground when it was rolled over would not be contacted as well by the cloth when compared to being clipped and placed into a jar of dislodging solution. The results showed that residues tended to be much higher with the turf-clipping technique. It can be seen in Table 2 that at all three sites the turf-clipping results were at least two times higher in concentration compared to the cloth-dosimeter technique results. Another way to look at this is to

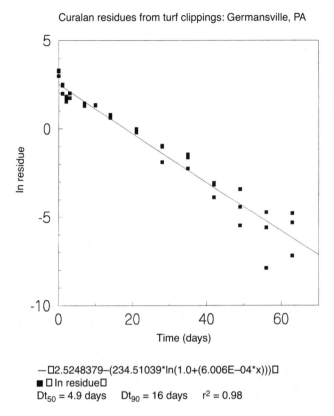

Figure 9 Decline curve for the Pennsylvania site using the turf-clipping method.

say that, in all three cases, the cloth dosimeters did not pick up as much residue as clipping the turf. Which value is correct? This is the question that the ORETF is trying to resolve with a study that encompasses the above-mentioned five techniques. That study will have not only the analytical results to determine the magnitude of the residues on the turf but will also include observations made by the investigators who actually did the different techniques. It is hoped that one of the methods will provide consistent results with little variation, thus indicating which technique can be done reproducibly by many people to provide a constant and reliable transfer coefficient.

From the results of this study, the conclusion can be drawn that the dislodgeable results tend to be variable from site to site and that the mean of the C_{envir} term will be dependent on the method of dislodging used on the turf. What is important is the magnitude of the final calculated transfer coefficient, which is also very dependent on the task that was done when generating the D_{pot} results. The differences produced by the different techniques for collecting the exposure data will affect the risk assessments performed using the data.

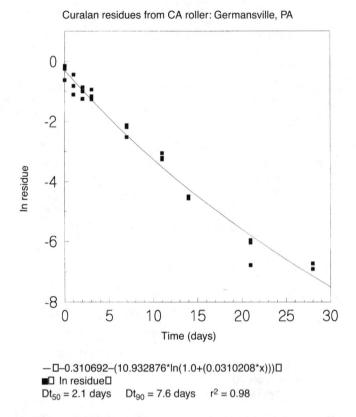

Curalan residues from CA roller: Germansville, PA

—☐–0.310692–(10.932876*ln(1.0+(0.0310208*x)))☐
■☐ ln residue☐
Dt_{50} = 2.1 days Dt_{90} = 7.6 days r^2 = 0.98

Figure 10 Decline curve for the Pennsylvania site using the California roller method.

Table 1 Dt_{50} and Dt_{90} Results After the Fourth Application

State	Turf clipping results (days)		California roller results (days)	
	Dt_{50}	Dt_{90}	Dt_{50}	Dt_{90}
California	2.9	11	1.5	5.9
Florida	1.1	3.8	9.5[a]	1.5
Pennsylvania	4.9	16	2.1	8.6

[a] Hours.

Table 2 0 Day Residue Results[a]

State	Turf clipping results ($\mu g/cm^2$)	California roller results ($\mu g/cm^2$)
California	10.6	1.24
Florida	3.75	0.645
Pennsylvania	24.5	0.725

[a] Conversion of $\mu g/mL$ to $\mu g/cm^2$ requires knowledge of the surface area that the turf sample represents. Knowing the amount of leaf area the clipped sample represents and the amount of AOT solution used to dislodge the sample gives the conversion relationship of mL AOT to leaf area sampled or mL of AOT sample to cm^2 of leaf.

Reference

Gustafson, D.I. and Holden, L.R. (1990) Nonlinear pesticide dissipation in soil: a new model based on spatial variability, *Environ. Sci. Technol.*, 24(7):1032–1038.

chapter eleven

Chemistry, GLP Standards, and worker safety

D.F. Hill and P. Swidersky

Contents

Analytical chemistry is a critical component of worker safety, re-entry, and other related studies intended to assess the risk to humans during and subsequent to pesticide applications. The analytical aspect takes on added significance when such studies are intended for submission to the U.S. Environmental Protection Agency and/or other regulatory authorities and are thus required to be conducted according to the Federal Insecticide, Fungicide and Rodenticide Act (FIFRA) Good Laboratory Practice (GLP) Standards, or their equivalent. This presentation will address test, control, and reference substance characterization, use-dilution (tank mix) verification, and specimen (exposure matrix sample) analyses from the perspective of GLP Standards requirements.

When one considers analytical chemistry and its general relationship to testing conducted according to the FIFRA GLP Standards,[1] there are three main study elements that need to be considered:

- Test, control, and reference substance characterization
- Test, control, and reference substance mixtures with carrier
- Specimen analyses

These three elements or phases certainly apply to worker safety and re-entry-related studies as much as for the more generally established and recognized toxicological/environmental effects testing, whether we are referring to inhalation/dermal exposure assessments or dislodgeable residue studies. The one analytical component often encountered in toxicology testing, but not generally included in worker safety studies, is "clinical chemistries," although blood and/or urine measurements, such as for cholinesterase inhibition, are a requirement of some re-entry-related studies.

Test, control, and reference (TCR) substances

In all GLP-applicable studies, the test substance is considered the actual chemical, mixture, or biological agent that is the basis of the study. With respect to pesticides regulated under FIFRA, the test substance is usually the subject of the marketing or research permit, but in some instances the test substance might be a metabolite, degradation product, by-product impurity, or radio-labeled compound. For most of the studies of concern to this symposium, the test substance will likely consist of a typical formulation of the pesticide that is the subject of the marketing permit or other registration requirement.

Worker safety studies are not likely to normally include a control substance (i.e., a material used in the study to serve as basis of comparison with the test substance). However, if a control substance is included as a treatment group, then it must: (1) be fully characterized as to its identity, purity (or strength), and stability (and solubility, if appropriate); (2) be appropriately tested in mixtures with any carrier used; and (3) meet all the other GLP recordkeeping, labeling, and storage requirements, as specified for the test substance. There is some regulatory relief here, however, in that water, by definition, is excluded from being considered a control substance, and vehicles (those substances added to enhance solubilization or dispersion of the test substance) are addressed separately in the FIFRA GLP Standards.

It is largely without debate that the term "reference substance" includes those chemical and biological materials that are used as standards in a study (i.e., those materials used for response comparison purposes, such as analytical reference standards). Normally, one thinks of reference substances as just referring to specimen (sometimes referred to as "matrix samples") analyses, but the characterization of test, control, and reference substances (see further discussion) itself will almost always require additional reference substances to determine purity or strength, as well as stability.

Some GLP professionals consider positive control substances (i.e., those materials intended to validate responsiveness of the test system to a chemical or biological challenge) to also be reference substances, as opposed to control substances; however, the point is somewhat moot as the FIFRA GLP Standards characterization and handling requirements are exactly the same for both control and reference substances. As with test and control substances, reference substances must be characterized (usually determination of purity and stability) according to the abbreviated GLP Standards (see next section).

With respect to worker safety and re-entry studies, reference substances are necessary to assay the test substance (and, if applicable, any control substance) and determine its stability and for the analyses of specimens collected in the study. Specimens may include plant material (dislodgeable residues), adsorbent media (inhalation), or clothing/dosimeter materials collected during a worker safety study to assess exposure. If biomonitoring is involved, blood and/or urine specimens may be analyzed against reference substances of known purity.

Characterization

The FIFRA GLP Standards require that all test, control, and reference substances that are employed in regulated studies be "characterized" before their actual use in such studies. The extent of "characterization" required is not spelled out in the GLP Standards, nor is there much in the way of even helpful hints. The FIFRA testing guidelines and other pesticide registration documents provide some guidance; however, it is largely left to the study management (in reality, usually the study director) as to what constitutes an adequate level of compliance and technical sufficiency in regard to this requirement. At a minimum, for any type of GLP study, the purity or strength of the test, control, and/or reference substances used should be verified analytically for active ingredient concentration or potency.

Some relief has been put forth, at least by the U.S. EPA, in that a GLP advisory[2] has been developed and distributed which states that the characterization of test, control, and reference substances used in GLP studies only needs to be accomplished according to the abbreviated form of the FIFRA GLP Standards, as given in 40 CFR 160.135. These requirements, which spell out limited GLP applicability to several basic product-testing requirements such as melting point, odor, viscosity, etc., preclude the need to characterize GLP reference substances by other fully GLP-characterized reference substances, thus avoiding the prospect of an endless analytical string of GLP characterizations requiring further GLP-characterized reference substances.

Test, control, and reference substances used in worker safety and re-entry testing also must be verified for their shelf-life stability prior to (or concomitantly with) use in a particular study. This means assuring stability of the substance for an extended period under controlled conditions or, if the material turns out to be unstable, determining what its useful life actually is. This stability testing does not have to be performed on exactly the same batch as that which will be used in the study of interest; however, the stability evaluation does have to be performed under the full GLP Standards requirements. This stability verification usually involves using standard analytical or bioassay procedures to determine periodically the active ingredient concentration in test substances stored under rigidly controlled conditions.

The stability for most test and reference substances that are active ingredients in experimental or commercial pesticides will already have been determined in response to the testing registration requirement given in the Product Chemistry Pesticide Assessment Guidelines,[3] Subdivision D. The FIFRA GLP Standards requirements for determination of stability concomitantly with a study require this stability testing to be conducted according to an approved standard operating procedure (SOP). This is often the only meaningful approach for long-term toxicological studies.

Test, control, and reference substance mixtures with carrier

The GLP requirements for mixtures with carrier (40 CFR 160.113) were originally intended to address problems associated with the incorporation of test and control substances into feed, water, and other media for toxicology studies; however, now these same requirements pertain to all other GLP-required studies, including those pertaining to re-entry and worker safety. The requirement involves: (1) substantiation of test, control, and reference substance concentration through periodic analyses; (2) verification of homogeneity; and (3) determination of stability and, if applicable, also solubility. All of these requirements usually require chemical analysis, although bioassay may be necessary for microbial pesticides.

Fulfilling this GLP requirement in an effective and meaningful fashion has proven to be quite problematic for field studies, including those related to worker safety and re-entry when tank mixes (use dilutions) using water as the carrier are involved. The lack of validated or proven sampling procedures, analytical problems associated with emulsion and suspensions, and the difficulties associated with interpretation of achieved analytical results all conspire against purposeful implementation of this aspect of the FIFRA GLP Standards requirement. The logistics and general difficulty of being able to obtain timely analyses (from the standpoints of both chemical stability and prompt feedback of analytical results) must also be considered detrimental factors in achieving meaningful compliance with the GLP requirements.

Although laboratory data regarding homogeneity, stability, effects of adjuvants, etc. may be available from the sponsor, this information is frequently not made available at the site of test substance application. A GLP deviation listed in the compliance statement may be the only realistic alternative without resorting to heroic effort; however, it must be kept in mind that use of this means of resolution needs to be a defensible and legitimate hardship (and be fully documented by the study director) and not just an expeditious means to avoid complying directly with the GLP requirement.

Specimen analysis

The collection and analysis of specimens (sometimes referred to as "matrix samples") are certainly of paramount significance to all worker safety studies, as the residue level or concentration of the pesticide (and/or related materials) represents the end-product or potential level of exposure, whether the specimens are vegetation, biomonitoring media (blood/urine), dosimeter patches, articles of clothing, or other adsorption media. It is somewhat ironic, considering the importance of specimen analyses and the subsequent results, that the GLP Standards do not directly address this aspect, except for specimen labeling and a non-specific SOP requirement for collection and identification. However, indirectly, the full force of facilities, equipment, personnel, protocol (methodology), and recordkeeping requirements all apply equally to specimen analysis as to other aspects of the study.

One aspect of specimen analysis that often occurs and should be highlighted is the situation that arises when a study has been initiated (protocol has been signed), but the analytical procedure has not yet been determined or worked out, or perhaps has not been fully validated by the performing laboratory. In this case, the approved protocol should fully describe the situation, and once the method has been developed and/or validated an approved protocol amendment should be issued, thus formalizing the inclusion of the analytical methodology. Likewise, during the validation process or during the study itself, if there is an analytical method modification then the protocol also needs to be formally amended.

It is recommended that the protocol itself contain language that allows for minor modifications of an analytical method or procedure without necessitating an amendment (or a deviation); for example, "Minor modifications in instrumental parameters and/or adjustments in technique may be made in the method during specimen analysis to enhance overall efficiency or the sensitivity, specificity, or selectivity of analyte response."

Method stages

Because several terms related to analysis have been used in this presentation without further explanation, it is worthwhile to elaborate briefly on the "method cycle" and relate the components in general terms to the GLP requirements. The three components of the cycle may be considered, in brief, to consist of:

- Method development
- Method validation
- Method application

These three analytical phases pertain to worker safety and re-entry testing as much as to other types of GLP studies, and, from a scientific soundness perspective, the method cycle also applies to test, control, and reference substance characterization (and mixtures with carriers).

Generally, method development, which can range from simply locating a method in the scientific literature to a full-blown laboratory research project, is considered to be research and development or exploratory in nature and thus not subject to the FIFRA GLP Standards requirements. However, it is strongly recommended that raw data associated with method development efforts be complete and recorded according to the GLP Standards. Also, all data and records should be fully attributable, organized, and retained — preferably retained in the GLP archives but separate from associated study files. If, for some reason, the method development phase has been included in a formal GLP study protocol, then this exploratory work also has to be conducted according to the full GLP Standards, although, if this inclusion was not intended and is detected in time, the method development requirement can be properly removed by an approved amendment.

Method validation, on the other hand, is normally considered part of the study in which the method will subsequently be used or consists of a separate defined study unto itself; as such, it is normally required to be accomplished under GLP purview. There is, however, some confusion in some circles as to exactly what is meant by analytical method validation. Some chemists describe it as adaptation of one method from one type of matrix for use with another using basically the same or similar analytical approach. Others take a more strict interpretation and define validation as simply demonstration of the ability to achieve satisfactory results using a published procedure in one's own laboratory setting. Often, validation incorporates both interpretations.

Regardless of the objective, the study director and advisory staff will need to determine the scope of validation prior to use of the method in specimen analyses. Criteria for acceptance of the method or for establishment of laboratory capability should also be established and documented *prior to* the validation process.

There has also been some difference of opinion as to what level of method reliability should be sought or achieved. The following list is provided to note at least some of the major factors that should be addressed during the validation process:

- Developing or locating a suitable method
- Obtaining adequate reference substance(s)
- Verifying procedural capability and adequacy
- Establishing a method detection limit (MDL)
- Verifying detector response linearity
- Identifying potential matrix effects on detector response
- Identifying potential interferences
- Establishing achievable accuracy and precision

- Identifying potential presence of metabolites/degradates/by-products
- Independent quality assurance assessment of validation effort
- Preparing a written report

If for some reason the method validation process is not a GLP study, or a component thereof, the laboratory should adhere to the same data recording and retention principles as described for method development.

With respect to method application, once validation has been satisfactorily completed, there is little question that use of the analytical method in worker safety and re-entry studies falls under the full requirements of the GLP Standards. In addition, there should be an adequate level of quality control measurements taken in conjunction with the specimens so as to provide for a meaningful assessment of accuracy and precision, as well as verification of freedom from artifactual interferences. Along with these measurements there needs to be reasonably rigid data acceptance criteria in place (usually established during validation) which are consistently applied during the course of the specimen analytical phase of the study.

References

1. Final Rule for Good Laboratory Practice Standards under the Federal Insecticide, Fungicide and Rodenticide Act, Code of Federal Regulations, Title 40, Part 160.
2. GLP Standards Advisories No. 1–75, U.S. Environmental Protection Agency, Office of Enforcement and Compliance Monitoring, Washington, D.C., 1989–1996.
3. Pesticide Assessment Guidelines, Subdivision D, Product Chemistry, U.S. Environmental Protection Agency, Office of Pesticide Programs, Washington, D.C., October, 1982.

Index

A

abamectin, 120, 124, 127, 128, 129
absorbed daily dose (ADD), 98, 99,
101, 102, 103, 104, 105
of chlorpyrifos, 104
acceptable operator exposure level
(AOEL), 92, 94
acetylcholine
accumulation of, 3
hydrolytic cleavage of, 3
acetylcholinesterase (ACHE), 3
inhibition of, by carbamates, 6, 7
inhibition of, by organo-
phosphorus pesticides, 3, 4, 5
ADD. *See* absorbed daily dose
air sampling, 23, 24, 26, 29, 51–52, 56,
57, 60, 68, 78
for cyromazine 75 WP, 88, 93
air-blast applicators, 22, 26, 28, 39,
41, 43
aldrin, 12, 13
alkylphosphates
detectable in urine, 6
organophosphorus pesticides,
and, 5
analyses, types of, 37–38
analytical chemistry, re-entry
studies and, 153

AOEL. *See* acceptable operator
exposure level
applicators, of pesticides, 22, 26, 27,
28, 31, 32, 38, 42, 43, 46, 65, 66,
67, 71, 72, 76, 77, 78, 79
atrazine, 14
mercapturic acid conjugate of, 14
metabolites of, 14–15
azinphos-methyl, 78

B

benzene hexachloride isomers, 12
biomarkers, 2, 16
carbon disulfide as, 8
for exposure to carbamates, 8
for exposure to organochlorine,
14
for exposure to
organophosphorus
pesticides, 4, 7
for exposure to pesticides, 9
for exposure to pyrethroids, 13
for pentachlorophenol exposure,
15
macromolecular adducts as, 16
urinary porphyrin pattern as, 13
biomonitoring. *See* monitoring:
biological

HARPER ADAMS UNIVERSITY
LIBRARY
COLLEGE